MEAN VALUE AND CORRELATION PROBLEMS CONNECTED WITH THE MOTION OF SMALL PARTICLES SUSPENDED IN A TURBULENT FLUID

N.V. VAN DE GARDE & CO'S DRUKKERIJ, ZALTBOMMEL

Mean Value and Correlation Problems connected with the Motion of Small Particles suspended in a turbulent fluid

PROEFSCHRIFT

TER VERKRIJGING VAN DE GRAAD VAN DOCTOR IN DE TECHNISCHE WETENSCHAP AAN DE TECHNISCHE HOGESCHOOL TE DELFT, OP GEZAG VAN DE RECTOR MAGNIFICUS PROF. J. M. TIENSTRA, HOOGLERAAR IN DE AFDELING DER WEG- EN WATERBOUWKUNDE, VOOR EEN COMMISSIE UIT DE SENAAT TE VERDEDIGEN OP DINSDAG 1 JULI 1947 TE 4 UUR,

DOOR

TCHEN CHAN-MOU

GEBOREN TE CHEKIANG

CHINA

Springer-Science+Business Media, B.V.

1947

ISBN 978-94-017-5737-9 ISBN 978-94-017-6101-7 (eBook)
DOI 10.1007/978-94-017-6101-7

Dit proefschrift en de stellingen zijn goedgekeurd door de promotor
PROF. DR. J. M. BURGERS

AAN MIJN OUDERS

CONTENTS

INTRODUCTION

The object of the present thesis is to elucidate a number of problems connected with the diffusion of irregularly moving particles, such as are found in particular when particles are suspended in a fluid in turbulent motion. For this purpose it was necessary to consider the general theory of dispersion phenomena in order to find out by which quantities the diffusion is characterized; then the values to be assigned to these quantities had to be derived from data characterizing the irregular motion of the liquid. A number of theorems on mean values and on correlation problems were necessary in order to make it possible to calculate the quantities involved.

The motion of small particles suspended in a liquid in turbulent motion in its theoretical aspect is a statistical problem, referring to the movements of a great number of particles, each of which is subjected to the irregular influences derived from the motion of the elements of volume of the fluid. The behaviour of an individual particle as a function of the time can be used as a starting point, but it would be impossible to consider every particle individually. It is necessary therefore to find the common properties of a multitude of particles; these can be obtained in part by considering the simultaneous behaviour of the particles belonging to a certain group; in part by following a single particle during a longer interval of time and calculating mean values over this interval.

In the first part of the investigation certain probability functions, referring to the displacements of particles or to their velocities at a given instant, will be used. These functions will be called *dispersion functions*. Such functions have been introduced already by various authors [1]; we shall base our investigation principally on the method developed by Kolmogoroff.

[1]) The application of probability and dispersion functions to kinetic problems has been started by v. Smoluchowski and by S. Chapman; see: M. v. Smoluchowski, *Bull. Ac. Cracovie*, (1913); *Göttinger Vorträge über kinetische Theorie der Materie*, 87, (1914), Leipzig; S. Chapman, *Proc. Roy. Soc.*, *A* **119**, 34, (1928).

Analytical investigations of the properties of such functions are due to Kolmogo-

In investigating diffusion phenomena by means of dispersion functions, it appeared useful to introduce two new points of view. In the first place the special case of dispersion in a system of constant density revealed the existence in such a case of a relation between the mean displacement and the mean square displacement, which makes itself felt by the appearence of a systematic flow of particles when the intensity of the turbulence is not a constant over the whole field. In view of the importance of this case the special name „*isomeric dispersion*" has been given to dispersion of this nature. Secondly it was found possible to introduce an extended form of the dispersion function which gives evidence concerning the distribution of the velocities at the end of the interval considered; with its aid a certain expression for the current of transportation could be formed in which this current appears as the resultant of a flow of dispersion in the direction of the gradient of turbulence, and a flow of concentration in the opposite direction. In isomeric dispersion these two flows compensate each other, so that in that case there is no resulting current.

The consideration of these problems forms the subject matter of Chapter 1, while in Chapter 2 some examples of dispersion functions are treated, illustrating the properties derived in Chapter 1.

In the 3rd Chapter the motion of a single particle is followed and time mean values are introduced. The most important time mean values are those of the square of the velocity, of the square of the length of the path, and of the product of the length of the path into the velocity at the endpoint. Taylor[1]) has shown the fundamental relation between the latter quantities and the average duration of correlation. In cases of steady mean flow the interval of time used in the calculation of time mean values can be taken of arbitrary length; difficulties, however, arise, when the flow is variable, or when particles wander into regions with a different intensity of turbulence. At a certain point of the investigation it becomes necessary therefore to return to mean values referring

---

roff, Hostinský; see: A. Kolmogoroff, *Math. Ann.*, **104**, 45–458, (1931); **108**, 149–160, (1933); B. Hostinský, *Mém. des Sc. Math.* Fasc. **52**; *Ann. de l'Inst. Henri Poincaré*, (1932) and (1937), and *Bull. Intern. de l'Acad. Tchèque des Sciences*, (1940).

A summary of Kolmogoroff's work with applications to ergodical systems has been given by: H. Gebelein, *Ann. der Physik* (5), **19**, 533, (1934).

[1]) G. I. Taylor, *Proc. Lond. Math. Soc.*, (2), **20**, 196–212, (1922).

to a great number of particles, or „ensemble mean values", and to investigate their relation to the time mean values. This forms the subject of the latter part of Chapter 3 (sections 3.3–3.34).

Next in Chapter 4, the relation between the motion of a single particle and that of the surrounding liquid is considered. For this purpose an equation is necessary, derived by Basset, by Boussinesq and by Oseen, which is the extension of Stokes' resistance formula to cases of variable motion. The particle is taken to be spherical, as for other shapes the relevant formulae have not been deduced or would be far too complicated. Some attention is given to the particular case of a periodic field, although this cannot represent the effect of the irregular fields occurring in reality; further it has been indicated in which way the equations can be extended to cases of motion where the exact linear resistance law fails.

On the basis of the results obtained in this Chapter, the quantities occurring in the investigation of time mean values as developed in Chapter 3 have been calculated in Chapter 5. As the behaviour of suspended particles presents a certain analogy with that of colloidal particles or the like in Brownian motion, in section 5.5 a comparison is made between certain formulae occurring in the two theories. It is found that the mean value of the product of the length of the path of a particle into the irregular force acting upon it, a quantity which is assumed to be zero in Langevin's theory of the Brownian motion, cannot be neglected under all circumstances, so that in the case of the particles considered here it must be taken into account.

In Chapter 6 the results obtained in the preceding Chapters are applied to the derivation of the equation governing the diffusion of the particles; in connection with this, various details of the reasoning are reviewed in order to point out where hypotheses had to be introduced or unsolved difficulties remain, in connection with the fact that in turbulent motion the intervals of time to be used in deducing mean values cannot be increased sufficiently in order to make them always considerably exceed the duration of correlation of the motion. An interesting problem in the diffusion of particles is connected with this matter, *viz.* whether in a case where the concentration of the particles is originally constant over the whole field, inequalities in the turbulent motion of the liquid can evoke differences in concentration, in such a way that the particles would tend to wander towards regions of weak turbulence. This problem, however, cannot be fully settled in view of the difficulties mentioned. Finally the dispersion equation obtained has been applied to some simple cases.

In so far as the formulae used depend upon the hydrodynamical equations, it has been assumed that the liquid or fluid in which the particles are suspended, is of constant density. This does not, however, exclude the application of the results to the dispersion of particles suspended in a gas, provided the velocities of the motion remain small in comparison with the velocity of sound in the gas.

Besides its interest from the theoretical side, the problem of the motion of suspended particles and of mean values and correlations connected with it, has a practical interest in investigations on the turbulence of a fluid; it is often attempted to deduce the character of the turbulent field from the measurements on the motion of suspended foreign matter used as an ,,indicator of turbulence''. Moreover, hydraulic engineers are interested in the phenomena of diffusion, presented by the particles in a given field of turbulent flow, as this leads to a knowledge of the distribution of solid matter in a current, which is of great importance. The reverse problem of the reaction of the particles upon the motion of the liquid, *e.g.* upon the transfer of momentum between adjacent layers of the latter, which may influence the velocity distribution of the main flow, will also be of interest in Hydraulics. This problem, however, has not been treated here and must be left for further researches.

Notation used for sections and subsections and numbering of the equations.

Sections and subsections will be numbered decimally, e.g. the second subsection of section 4 of Chapter 3 will be numbered 3.42. Equations are numbered consecutively afresh in every subsection (or section); references in the same subsection (or section) are made by mentioning the number of the equation (written between brackets) only; references to equations given in other subsections (or sections) will be preceded by the indication of the section or subsection. In this way the 3rd equation of subsection 3.42 will be referred to as 3.42(3).

CHAPTER 1

STATISTICAL CONSIDERATIONS CONCERNING THE DISPERSION OF PARTICLES IN IRREGULAR MOTION [1])

1.1. Definition of the dispersion function

In problems concerning the motion of particles as occurring in Mechanics and Physics, we can distinguish 2 types:

(1) problems relating to completely determinate systems;

(2) problems relating to irregular systems.

In the problems of the first type the motion of the particles is completely controlled by certain known equations, so that the whole course of the motion can be derived from prescribed initial conditions. It is evident that the motion can be calculated for increasing time as well as for decreasing time. It is the usual aim of Classical Mechanics to investigate problems of this type.

In the problems of the 2nd type, on the contrary, the motion of the particles cannot exactly be determined, for we never know the detailed initial conditions nor all the complicated irregular influences to which the particles may be subjected. Such motions are found when particles are subjected, for instance, to the irregular influences of mutual impacts or of impacts with the molecules of the medium in which they are embedded; examples are found in turbulent motion, Brownian motion, and the motion of particles considered in the kinetic theory of gases. For these motions it is only possible to calculate a distribution of the chances for the occurrence of particular positions and velocities of the particles at a later instant, assuming that certain initial conditions have been given. Such systems can be called irregular systems; in some publications

---

[1]) This chapter in the main follows the exposition given in a paper „Enige wiskundige betrekkingen welke een rol spelen in diffusieproblemen" by the present author, *Versl. Ned. Akad. v. Wetensch., afd. Natuurk.* **53**, 400—410, (1944). The treatment of the problems, however, has been extended in various points.

they are denoted also as *Stochastical systems.* It is the aim of Statistical Mechanics to investigate problems of this type.

We will take the simplest case where the position of a particle is characterized by one coordinate y only. Provisionally the velocity which the particle possesses at any instant will not be introduced as an explicit variable; an extension to the case where explicit attention is given to the velocities is deferred until section 1.7. It is assumed that at a given instant t_0 there are many particles with positions between y_0 and $y_0 + dy_0$. Let the number of these particles be:

$$n(t_0, y_0)dy_0, \tag{1}$$

where n will be called the particle density.

Let

$$p(t_0, y_0; t, y)dy \tag{2}$$

be the probability for a particle which at the instant t_0 started from the region $(y_0, y_0 + dy_0)$, to arrive in the region $(y, y + dy)$ at the instant t. The probability function p will be called *the probability function of displacement.* It controls the dispersion process of particles as due to the irregular motion, hence it can also be called *the dispersion function.*

The function p must satisfy the relation

$$\int_{-\infty}^{+\infty} dy \;\; p(t_0, y_0; t, y) = 1. \tag{3}$$

Here and in following formulae the limits of integration for convenience are given as $-\infty$, $+\infty$; the meaning of the formulae is that the integration is carried out over the whole available domain of y-values, which often may be of limited extent. Formula (3) expresses the condition that all the particles coming out of the region dy_0 must find their place in the totality of the elements dy forming the available domain of the variable y. — As will be explained in 1.2, the picture of dispersing particles can be extended in such a way that it can be applied to the dispersion of the elements of volume of a liquid in turbulent motion.

Let the number of particles which at the instant t are situated in the region $(y, y + dy)$ be $n(t, y)dy$. When at the instant t_0 the number of particles $n(t_0, y_0)dy_0$ is given for all intervals dy_0, it is evident that the number of particles $n(t, y)dy$ can be deduced from the relation

$$n(t, y) = \int_{-\infty}^{+\infty} dy_0 \; n(t_0, y_0) \; p(t_0, y_0; t, y). \tag{4}$$

Evidently

$$\int_{-\infty}^{+\infty} dy \; n(t, y) = \int_{-\infty}^{+\infty} dy_0 \, n(t_0, y_0). \tag{5}$$

When for the function n we choose such a unit that

$$\int_{-\infty}^{+\infty} dy_0 \; n(t_0, y_0) = 1; \qquad \int_{-\infty}^{+\infty} dy \; n(t, y) = 1,$$

then $n(t, y)$, which previously has been called the density of the particles, can be considered as the probability for the particles to be situated in the region $(y, y + dy)$ at the instant t.

1.11. Further properties of the dispersion function

An important problem in the theory of dispersion functions is to determine the minimum value to which the interval of time $t - t_0$ may decrease before it becomes impossible to assume that the disturbing actions (due to impacts and the like) experienced by the particles exercise sufficiently strong influences in order to allow the application of probability considerations. In our case where the initial velocity is not introduced as a separate parameter, the particles which are situated in dy_0 at the initial instant t_0 will possess unequal velocities, so that already immediately a dispersion may be expected. In general, however, we must assume that the interval $t - t_0$ shall be of sufficient length in order that every particle may have suffered a serious disturbance of its movement. This can be expressed in other words by requiring that $t - t_0$ shall exceed the duration (or better still the upper measure ϑ) of the correlation to be found in the course of the velocity function for a single particle, a quantity which will be introduced in Chapter 3 (sections 3.121 and 3.13).

In order to be able to deduce a set of differential equations Kolmogoroff has assumed that the dispersion function can be treated as an analytic function for arbitrarily small intervals of $t - t_0$[1]. There seems to be little objection to such a procedure

[1]) Compare: A. Kolmogoroff, „Über die analytischen Methoden in der Wahr-

when the functions to which it is applied vary sufficiently slowly with the time in order to make differential quotients a satisfactory approximation to difference quotients, calculated for an increase of time sufficiently exceeding the correlation measure mentioned above. However, as there may be certain cases in which some care is required in dealing with small intervals, we shall provisionally retain several equations in the form of difference equations with respect to the time, indicating the corresponding differential equations in those cases where this appears to be of importance.

It is in the nature of the phenomena of motion, whether regular or irregular, that we must expect

$$\lim_{t=t_0} p(t_0, y_0; t, y) = 0 \quad \text{for } y - y_0 \neq 0. \tag{1}$$

As nevertheless relation 1.1(3) must remain valid, it is necessary to assume that for $y = y_0$ the function $p(t_0, y_0; t, y)$ will take an integrable infinite value when t tends indefinitely to t_0:

$$\lim_{t=t_0} p(t_0, y_0; t, y) = \infty \quad \text{for } y - y_0 = 0. \tag{2}$$

In order to conform to the laws of diffusion phenomena, the dispersion function moreover must satisfy the condition

$$p(t_0, y_0; t, y) = \int_{-\infty}^{+\infty} dy'' \; p(t_0, y_0; t'', y'') \; p(t'', y''; t, y), \tag{3}$$

where $t_0 < t'' < t$. This equation guarantees that the diffusion process due to the irregular motion is integrable, which means that the dispersion of the particles in the interval of time $t - t_0$ as controled by the function $p(t_0, y_0; t, y)$ can be obtained by calculating first the dispersion in an interval of time $t'' - t_0$ and then in the adjoining interval $t - t''$, t'' being an arbitrary intermediary instant within the interval $t - t_0$. In principle it may be regarded as a selfevident relation, or as an equation defining $p(t_0, y_0; t, y)$, when the functions occurring under the integral sign

scheinlichkeitsrechnung", *Mathem. Ann.* **104**, 415, (1931); ,,Zur Theorie der stetigen zufälligen Prozesse", *ibid.* **108**, 149, (1933); *Annals of Mathem.* **35**, 116, (1934).

An outline of Kolmogoroff's formulae (with applications to problems concerning ergodical systems) is given by: H. Gebelein, *Ann. der Physik* (5), **19**, 533, (1934). Compare also: M. von Smoluchowski, *Ann. der Physik* (4), **48**, 1106, (1915); and H. C. Burger, *Versl. Kon. Akad. v. Wetensch.*, Amsterdam, **25**, 1106 and 1482, 1917.

have been given. Equation (3), however, becomes of importance when it is required that all three p-functions occurring in it shall be of the same form with respect to their arguments. It is possible to extend the condition by considering not one intermediate instant t'', but more such instants:

$$p(t_0, y_0; t, y) = \int_{-\infty}^{+\infty} dy'' \int_{-\infty}^{+\infty} dy''' \dots \int_{-\infty}^{+\infty} dy^{(n)}\ p(t_0, y_0; t'', y'') \, . \\ \cdot\, p(t'', y''; t''', y''') \dots p(t^{(n)}, y^{(n)}; t, y). \quad (4)$$

1.12. Mean values connected with the dispersion process

In all dispersion problems an important part is played by the mean values of the displacement and of its second and higher powers. We introduce the notation:

$$\overline{l^m} = \int_{-\infty}^{+\infty} dy\ (y - y_0)^m\ p(t_0, y_0; t, y). \quad (1)$$

The mean value of the displacement itself, $\overline{l}$, will be zero in the case of symmetrical dispersion; in other cases it can be expected to be of the order of the time interval $t - t_0$. The mean value of the square of the displacement, $\overline{l^2}$, for large values of the interval $t - t_0$ usually may be expected to increase proportionally with $(t - t_0)^2$ [1]). For small values of this interval (but still exceeding the correlation measure [2])), however, we must not expect it to vanish proportionally with $(t - t_0)^2$: in consequence of the circumstance that even for small values of $t - t_0$ there is a finite possibility for large (positive and negative) values of the displacement, $\overline{l^2}$ can be of the order of $t - t_0$ for small values of this interval. This can be seen when as example for p we take Gauss' function:

$$p = \frac{1}{\sqrt{\pi k(t - t_0)}}\, e^{-(y-y_0)^2/k(t-t_0)}, \quad (2)$$

k being a constant. — That the finite possibility for the occurrence of large values of the displacement need not affect the order of magnitude of $\overline{l}$, is due to the circumstance that large positive and

[1]) This is not always true; compare Chapter 2, equation 2.3(7).

[2]) See definition in 3.121 and 3.13.

large negative displacements may practically cancel each others contributions towards this quantity.

Gauss' function, which usually is considered as a typical example of a dispersion function, further has the property that the mean values of higher powers of the displacement, as $\overline{l^3}$ etc., for small intervals $t - t_0$ are small compared with $t - t_0$. Following Kolmogoroff we shall assume this to be a general property of the dispersion functions to be considered here. We must leave open the question whether there exist other types of dispersion functions which do not possess this property.

Kolmogoroff moreover assumes that the ratios:

$$\overline{l}/(t - t_0) \quad \text{and} \quad \overline{l^2}/(t - t_0)$$

approach to constant values when the interval $t - t_0$ is decreased indefinitely. It is not certain that this always will be the case; indeed when the interval $t - t_0$ is decreased below the correlation measure we must expect that the probability considerations which form the basis for the introduction of the dispersion function will not apply. Circumspection therefore is necessary when we wish to make use of such an assumption. In the case of Gauss' function it does apply, which is a consequence of the fact that this function is a limiting type for zero correlation measure.

1.13. Series development of the dispersion function

In Gauss' function the parameters t_0, y_0, t, y figure exclusively in the form of the differences $t - t_0$, $y - y_0$. In functions of general type, however, t_0, y_0 themselves (or in another representation, t, y) must also be present.

Let:

$$t - t_0 = \tau; \; y - y_0 = l, \tag{1}$$

then we can write the dispersion function $p(t_0, y_0; t, y)$ as a function P of $(t_0, y_0; \tau, l)$ or of $(t - \tau, y - l; \tau, l)$:

$$p(t_0, y_0; t, y) = P(t_0, y_0; \tau, l) = P(t - \tau, y - l; \tau, l). \tag{2}$$

This second mode of writing is useful when we want to express that a dispersion function varies more slowly with t_0, y_0 than with τ, l. In particular this will be the case when τ is small:

$$\partial P/\partial y_0 \ll \partial P/\partial l \qquad \text{for } \tau \text{ small.} \tag{3}$$

In order to make calculations with the aid of the functions p and P, it will be of importance to know whether they can be developed into a Taylor series. It will be assumed that when $t - t_0 = \tau$ is not a small quantity, the function p can be developed into a Taylor series with respect to the variable y_0, as well as to the variable y. For example we shall write (as will be used in 1.3 and 1.31):

$$p(t_0, y_0 + l; t, y) = p(t_0, y_0; t, y) + l\frac{\partial p}{\partial y_0} + \frac{l^2}{2}\frac{\partial^2 p}{dy_0^2} + \ldots \quad (4)$$

$$p(t_0, y_0; t, y - l) = p(t_0, y_0; t, y) - l\frac{\partial p}{\partial y} + \frac{l^2}{2}\frac{\partial^2 p}{\partial y^2} - \ldots \quad (5)$$

When $t - t_0 = \tau$ is a small quantity, then in connection with 1.11(2), the case may exist that $\partial p/\partial y_0$ in (4) resp. $\partial p/\partial y$ in (5) will be large and the series (4) resp. (5) will be very badly convergent or even may cease to be convergent. In this case it will be assumed that if the dispersion function is written in the form P [form. (2)], a development of P will still be possible into a Taylor series with respect to y:

$$P(t - \tau, y - l; \tau, l) = P(t - \tau, y; \tau, l) - l\frac{\partial P}{\partial y} + \frac{l^2}{2}\frac{\partial^2 P}{\partial y^2} - \ldots \quad (6)$$

This development of P into a Taylor series with respect to y means a development of $p(t_0, y_0; t, y)$ into a series proceeding simultaneously with respect to y_0 and to y, with equal increments of both variables.

Under the conditions stated in 1.1—1.13 it appears that the function p will satisfy two equations with derivatives of the second order with respect to y, which have originally been given by Kolmogoroff in the form of partial differential equations. The second one of these partial differential equations, which is the most important one, is derived by Kolmogoroff with the aid of a process which does not seem to be quite exact. In the following pages, independently from Kolmogoroff's methods, we will first develop some formulae concerning mean values, and then return to these partial differential equations.

By a similar process as used in the derivation of the equations for p we can obtain an equation for the function n. This equation appears to be of a similar type as one derived by Fokker and by P l a n c k in an investigation connected with Quantum Theory.

It is to be remarked that along with the above mentioned dispersion function which refers to increasing time, the inverse problem might be formulated: we could ask from which preceding distribution an observed distribution of particles may have originated. Such „inverse dispersion functions", however, present special difficulties.

1.2. Problems with constant density

An important subdivision of the systems considered above is obtained in the case where the quantity n, which represents the local density of the particles, is a constant, independent of t as well as of y. In this case we shall speak of *isomeric dispersion* and correspondingly of *isomeric dispersion functions*. We can give an illustration of such a case by supposing that in a two-dimensional field occupied by a liquid of constant density, there exists an irregular motion, the components v and w of which are functions of y, z and of the time t, while satisfying the equation of continuity. Such a motion will produce a never ending change of position of the elements of volume of the liquid. Schematically its effect can be represented by imagining the whole field to be divided into square elements of equal area, and supposing that these elements are continually permuted over the field in such a way that at each permutation not only contiguous elements may interchange places, but also elements at arbitrary distances from each other. When now we restrict to the registration of the displacements in the direction of y, in this case again we obtain a system to which the probability formulae of the preceding sections can be applied, as for any element with a given value of the coordinate y_0 at the instant t_0 it still is a matter of pure chance what will be its displacement in the interval from t_0 to t.

In the case of a real liquid in turbulent motion the elements of area displaced by the irregular movements will be unequal in form and magnitude. The case could be reduced to the one considered above by subdividing the actual elements in such a way that equal squares or cubes would be obtained. However, we will not go into an investigation of the implications which may be connected with such a subdivision, and in the following considerations we will assume that all elements are equal in form and magnitude so that they can take the place of particles. It is necessary of course to assume that these permutations be performed so quickly that in a small interval of time $\tau = t - t_0$ a certain diffusion already takes place, in order to comply with the considerations developed in 1.1, 1.11 and 1.12. In such a system the number of particles in an interval dy_0 will be equal to dy_0 multiplied by a constant factor depending upon the breadth in the z-direction. The system thus obtained can serve as an image of the diffusion of the volume elements of an incompressible fluid or other substratum in turbulent motion.

From 1.1(4) there now follows

$$\int_{-\infty}^{+\infty} dy_0 \, p(t_0, y_0; t, y) = 1. \tag{1}$$

Hence the integral of an isomeric dispersion function with respect to dy_0 gives the value 1, just as did the integral with respect to dy in 1.1(3).

From the general formula 1.1(4) in which n was not supposed to be a constant, we read how the total number of particles which are situated in dy at the instant t is composed of contributions coming from everywhere in the whole region. Therefore we may consider

$$p'(t_0, y_0; t, y) = \frac{n(t_0, y_0)}{n(t, y)} p(t_0, y_0; t, y) \tag{2}$$

as the probability for a particle to have started from dy_0 at the instant t_0, given its situation dy at the instant t. In a certain sense the function p' represents a kind of inverse function of the dispersion function p. However, p' is dependent upon the particles really present and has no independent meaning such as was the case for the function p. In order to prevent any possible confusion with real inverse dispersion functions as mentioned in 1.13 we will provisorily call p' the „*retrograde dispersion function*".

When we restrict to the case of isomeric dispersion, n disappears from the above formula and we obtain

$$p'(t_0, y_0; t, y) = p(t_0, y_0; t, y). \tag{3}$$

Hence an isomeric dispersion function is identical with its own retrograde function.

1.21. Quantities which are characteristic for the mean displacement of the particles

In the present section we write $t - t_0 = \tau$, $y - y_0 = l$, as was done already in 1.13(1); for convenience we omit the indices in $t_0\, y_0$. Hence the dispersion function for particles coming from the region between y and $y + dy$ at the instant t — for the sake of brevity we may speak of the particles starting from (t, y) — will be written

$$p(t, y; t + \tau, y + l) = P(t, y; \tau, l). \tag{1}$$

In all following integrals the limits are to be read $-\infty$, $+\infty$, which means that they are extended over the whole domain available for y or, what comes to the same, for l. Now we introduce the quantities:

$$\overline{l} = \int dl\, l\, P(t, y; \tau, l), \tag{2}$$

$$\overline{l^2} = \int dl\, l^2\, P(t, y; \tau, l), \tag{3}$$

$$\bar{l} = \frac{1}{n(t, y)} \int dl\, l\, n(t - \tau, y - l)\, P(t - \tau, y - l; \tau, l), \tag{4}$$

$$\bar{l^2} = \frac{1}{n(t, y)} \int dl\, l^2\, n(t - \tau, y - l)\, P(t - \tau, y - l; \tau, l). \tag{5}$$

Here $\overline{l}$ determines the mean value of the displacement in the interval of time $t \rightarrow t + \tau$ of the particles starting from (t, y). In the same way $\overline{l^2}$ gives the mean square of the displacements On the contrary $\bar{l}$ gives the mean value of the displacement of the particles at t, y in the preceding interval $t - \tau \rightarrow t$. While $\overline{l}$ and $\overline{l^2}$ refer to the dispersion process starting from y, the quantity $\bar{l}$ may be considered as referring to the „concentration process" towards y; hence we shall call $\overline{l}$ *mean displacement of dispersion* and $\bar{l}$ *mean displacement of concentration.* The corresponding mean square displacement is represented by $\bar{l^2}$.

By introducing a „retrograde dispersion function"

$$P'(t, y; \tau, l) = \frac{n(t, y)}{n(t + \tau, y + l)} P(t, y; \tau, l), \tag{6}$$

form. (2)—(5) also may be written:

$$\bar{l}(t, y) = \frac{1}{n(t, y)} \int dl\, l\, n(t + \tau, y + l)\, P'(t, y; \tau, l), \tag{7}$$

$$\overline{l^2}(t, y) = \frac{1}{n(t, y)} \int dl\, l^2\, n(t + \tau, y + l)\, P'(t, y; \tau, l), \qquad (8)$$

$$\bar{l}(t, y) = \int dl\, l\, P'(t - \tau, y - l; \tau, l), \qquad (9)$$

$$\overline{l^2}(t, y) = \int dl\, l^2\, P'(t - \tau, y - l; \tau, l). \qquad (10)$$

In the particular case of isomeric dispersion there are found some simple relations between $\bar{l}$, $\overline{l^2}$, and $\bar{l}$. In this case, as follows from 1.2(3), eqs. (4), (5) simplify to:

$$\left.\begin{aligned} \bar{l} &= \int dl\, l\, P(t - \tau, y - l; \tau, l) \\ \overline{l^2} &= \int dl\, l^2\, P(t - \tau, y - l; \tau, l) \end{aligned}\right\} (n \text{ constant}). \qquad \begin{aligned} (11) \\ (12) \end{aligned}$$

In all following deductions it is supposed that with increasing values of $|l|$ the function P decreases sufficiently quickly in order that terms with the factors l^3 etc. may be neglected in comparison with terms with the factors l or l^2. This requires of course that τ is small; compare the considerations developed in connection with eq. 1.12(1).

We start from eq. 1.2(1), which applies to isomeric functions and which can be written:

$$\int dl\, P(t - \tau, y - l; \tau, l) = 1. \qquad (13)$$

In accordance with 1.13(6) we develop P into a Taylor series, restricting to the terms of the orders l and l^2; we then obtain:

$$\int dl \left\{P(t - \tau, y; \tau, l) - l\frac{\partial P}{\partial y} + \frac{l^2}{2}\frac{\partial^2 P}{\partial y^2}\right\} = 1.$$

According to 1.1(3) the integral of the first term is equal to 1, independently of the value of τ. The integrals of the remaining terms taken together therefore must be equal to zero. Applying eqs. (2), (3) given above we arrive at the result:

$$-\frac{\partial \bar{l}}{\partial y} + \frac{\partial^2}{\partial y^2}\left(\frac{\overline{l^2}}{2}\right) = 0. \qquad (14)$$

In this way a relation is obtained between the mean displacement and the mean square displacement, for small τ, valid for isomeric dispersion. Integration with respect to y gives:

$$\overline{l} = \frac{\partial}{\partial y}\left(\frac{\overline{l^2}}{2}\right) + f(t_0, t). \tag{15}$$

The integration constant which has been written as $f(t_0, t)$ in this equation, in many cases can be put equal to zero, *e.g.* when the domain where the diffusion takes place gradually passes into a region in which there are no appreciable displacements of the particles. This requires the absence of any peculiar motion produced by the action of exterior forces such as *e.g.* gravity (the integration constant also could be eliminated by superposing a general translation with the appropriate velocity). For the present we shall assume that this is the case, intending to return to the more general case in 1.6; we thus have:

$$\overline{l} = \frac{\partial}{\partial y}\left(\frac{\overline{l^2}}{2}\right). \tag{16}$$

We next apply a similar development to form. (11) by means of which $\bar{l}$ was defined. This gives:

$$\bar{l} = \int dl\; l \left\{P(t-\tau, y; \tau, l) - l\frac{\partial P}{\partial y}\right\},$$

which reduces to:

$$\bar{l} = \overline{l} - \frac{\partial}{\partial y}\left(\overline{l^2}\right) = -\overline{l}. \tag{17}$$

In an analogous way it can be found that:

$$\bar{l^2} = \overline{l^2}, \tag{18}$$

where $\bar{l^2}$ is defined by (12). It is to be noted that in the derivation of these formulae no use has been made of the integral relation 1.11(3); hence they hold good independently of it.

Equation (17) also can be written:

$$\bar{l} = -\frac{\partial}{\partial y}\left(\frac{\overline{l^2}}{2}\right). \tag{17a}$$

Consider the case of a turbulent field where the turbulence increases with increasing values of y, so that:

$$\frac{\partial \overline{l^2}}{\partial y} > 0.$$

Elements diffusing out of a definite region dy will take ever greater movements when they are displaced in the positive direction. Hence they will have greater chances to be dispersed further away than elements displaced in the opposite direction, and at the end of a small interval of time there will result a mean displacement $\overset{\sqcap}{l}$ in the positive direction proportional to the gradient of the turbulence. This is the meaning of the relation (16). The fact that even in the absence of a peculiar motion produced by exterior forces, there can appear a mean displacement of diffusion $\overset{\sqcap}{l}$ of the same order of magnitude as the time interval $t - t_0$, is an important feature of a field with inhomogeneous turbulence. Although deduced here for the case of isomeric dispersion, it may be found likewise in other cases of inhomogeneous fields.

By the departure of the dispersed elements the region dy would gradually be emptied. In the case of isomeric dispersion, however, it must be refilled at the same time by other elements. The mean displacement of the elements diffusing into dy is given by the quantity $\overline{l}$, the value of which is determined by relation (17a). The fact that for isomeric dispersion the ,,mean displacement of dispersion'' $\overset{\sqcap}{l}$ (the mean displacement of the elements diffusing *out of dy*) and the ,,mean displacement of concentration'' $\overline{l}$ (the mean displacement of the elements diffusing *into dy*) are equal and opposite, ensures equilibrium. We consequently may consider it as a characteristic of isomeric diffusion that simultaneously with a flow of dispersion there is a flow of concentration. An example illustrating the relation between $\overset{\sqcap}{l}$ and $\overset{\sqcap}{l^2}$ for isomeric dispersion will be given in Chapter 2.

For certain purposes it is convenient to denote the quantity $\overset{\sqcap}{l}$ (or the corresponding mean velocity $\overset{\sqcap}{l}/\tau$), which has already been called the ,,mean displacement of dispersion'', as the *systematic motion* (resp. *systematic velocity*), independently whether its presence is due to inhomogeneous turbulence or to the action of exterior forces. When exterior forces are active, we shall call their contribution to $\overset{\sqcap}{l}$ (resp. $\overset{\sqcap}{l}/\tau$) the *peculiar motion*. In the phenomenological treatment given here it is only the systematic motion which can be distinguished.

1.3. Derivation of two equations satisfied by the dispersion function

We will now make use of the property expressed by 1.11(3)

$$p(t_0, y_0; t, y) = \int dy'' \ p(t_0, y_0; t'', y'') \ p(t'', y''; t, y), \qquad (1)$$

with $t_0 < t'' < t$, by which, when added to the condition that the function p shall always be of the same form with respect to its arguments and can be subjected to differentiation with respect to them, the dispersion function is subjected to rather severe conditions. In this expression let $t'' - t_0 = \tau$ represent a small quantity. When $t - t_0$ is not particularly small, this neither will be the case with $t - t'' = t - t_0 - \tau$. When we write $y'' = y_0 + l$ and take into account what has been assumed in 1.11, we may develop

$$p(t'', y''; t, y) = p(t'', y_0 + l; t, y)$$

into a Taylor series:

$$p(t'', y_0 + l; t, y) - p(t'', y_0; t, y) + l \frac{\partial p}{\partial y_0} + \frac{l^2}{2} \frac{\partial^2 p}{\partial y_0^2} + \ldots .$$

Substituting this value into (1), and applying form. 1.1(3) and 1.12(1) we obtain:

$$p(t_0, y_0; t, y) = \int dl \, p(t_0, y_0; t'', y_0 + l) \, p(t'', y_0 + l; t, y) =$$

$$= \int dl \, p(t_0, y_0; t'', y_0 + l) \left\{ p(t'', y_0; t, y) + l \frac{\partial p}{\partial y_0} + \frac{l^2}{2} \frac{\partial^2 p}{\partial y_0^2} \right\} =$$

$$= p(t'', y_0; t, y) + \overline{l}(t_0, y_0) \frac{\partial p}{\partial y_0} + \frac{\overline{l^2}}{2} \frac{\partial^2 p}{\partial y_0^2} .$$

Now we shall write:

$$p(t'', y_0; t, y) - p(t_0, y_0; t, y) = \delta_0 p(t_0, y_0; t, y).$$

Then we find:

$$\delta_0 \, p(t_0, y_0; t, y) = -\overline{l}(t_0, y_0) \frac{\partial p}{\partial y_0} - \frac{\overline{l^2}}{2} \frac{\partial^2 p}{\partial y_0^2} . \qquad (2)$$

This is the equation for the change of p with initial time t_0.

In the second place we put $t'' = t - \tau$, $y'' = y - l$, and write (1) in the form:

$$p(t_0, y_0; t, y) = \int dl \ p(t_0, y_0; t'', y - l) \ P(t'', y - l; \tau, l). \qquad (3)$$

As now $t'' - t_0$ is not a small quantity, we may develop the product $p(t_0, y_0; t'', y - l)\ P(t'', y - l; \tau, l)$, according to what has been assumed in 1.11, into the series

$$p(t_0, y_0; t'', y)\ P(t'', y; \tau, l) - l\frac{\partial}{\partial y}(pP) + \frac{l^2}{2}\frac{\partial^2}{\partial y^2}(pP).$$

Substituting this into (3) we obtain, after integration with respect to dl:

$$p(t_0, y_0; t, y) = p(t_0, y_0; t'', y) - \frac{\partial}{\partial y}[\overline{l}(t'', y)\, p] + \frac{\partial^2}{\partial y^2}\left[\frac{\overline{l^2}}{2}\, p\right].$$

Hence if we write

$$p(t_0, y_0; t, y) - p(t_0, y_0; t'', y) = \delta\, p(t_0, y_0; t, y),$$

there follows:

$$\delta\, p(t_0, y_0; t, y) = -\frac{\partial}{\partial y}[\overline{l}(t, y)\, p] + \frac{\partial^2}{\partial y^2}\left[\frac{\overline{l^2}}{2}\, p\right], \tag{4}$$

where on account of the smallness of $\tau = t - t''$, we have written $\overline{l}(t, y)$ instead of $\overline{l}(t'', y)$. This is the equation for the change of p with final time t.

In the derivation of (2) and (4) use has been made exclusively of formulae which are valid for variable n as well as for constant n; hence it follows that the equations hold good for both cases.

When now we introduce Kolmogoroff's assumptions that for indefinitely decreasing values of τ the ratio's $\overline{l}/\tau$ and $\overline{l^2}/\tau$ assume constant values, we can replace eqs. (2) and (4), after division by τ, by the partial differential equations:

$$\frac{\partial p(t_0, y_0; t, y)}{\partial t_0} = -\frac{\overline{l}}{\tau}(t_0, y_0)\frac{\partial p}{\partial y_0} - \frac{\overline{l^2}}{2\tau}\frac{\partial^2 p}{\partial y_0^2}; \tag{5}$$

$$\frac{\partial p(t_0, y_0; t, y)}{\partial t} = -\frac{\partial}{\partial y}\left[\frac{\overline{l}}{\tau}(t, y)\, p\right] + \frac{\partial^2}{\partial y^2}\left[\frac{\overline{l^2}}{2\tau}\, p\right]. \tag{6}$$

1.31. Equations for isomeric dispersion functions

When we restrict to isomeric functions as considered in 1.2 we can apply 1.2(1), 1.21(11) and 1.21(12) to the transformation of the integral equation 1.11(3). By putting $t'' = t_0 + \tau$, $y'' = y_0 + l$ this equation can be written:

$$p(t_0, y_0; t, y) = \int dl\, P(t_0, y_0; \tau, l)\ p(t_0 + \tau, y_0 + l; t, y). \tag{1}$$

As τ is small and $t - t_0 - \tau$ remains finite, the product

$$P(t_0, y_0; \tau, l)\, p(t_0 + \tau, y_0 + l; t, y)$$

can be developed into a Taylor series according to the considerations already applied in 1.21, and we can write:

$$p(t_0, y_0; t, y) = \int dl \Big\{ P(t_0, y_0 - l; \tau, l)\; p(t_0 + \tau, y_0; t, y) + + l \frac{\partial}{\partial y_0}(pP) + \frac{l^2}{2}\frac{\partial^2}{\partial y_0^2}(pP)\Big\} =$$

$$= p(t_0 + \tau, y_0; t, y) + \frac{\partial}{\partial y_0}[\bar{l}(t'', y_0)\, p] + \frac{\partial^2}{\partial y_0^2}\left[\frac{\overline{l^2}}{2} p\right].$$

Hence we obtain:

$$\delta_0\, p(t_0, y_0; t, y) = -\frac{\partial}{\partial y_0}[\bar{l}(t_0, y_0)\, p] - \frac{\partial^2}{\partial y_0^2}\left[\frac{\overline{l^2}}{2} p\right], \tag{2}$$

where $\bar{l}(t_0, y_0)$ has been substituted for $\bar{l}(t'', y_0)$.

On the other hand when we put $t'' = t - \tau$, $y'' = y - l$, the integral equation 1.11(3) can be written as:

$$p(t_0, y_0; t, y) = \int dl\; p(t_0, y_0; t - \tau, y - l)\; p(t - \tau, y - l; t, y). \tag{3}$$

As $t - \tau - t_0$ is not small the first factor can be developed into the series [see 1.13(5)]:

$$p(t_0, y_0; t - \tau, y) - l\frac{\partial p}{\partial y} + \frac{l^2}{2}\frac{\partial^2 p}{\partial y^2}.$$

Substituting this value into (3) and applying form. 1.2(1), 1.21(11), 1.21(12) we obtain:

$$p(t_0, y_0; t, y) = p(t_0, y_0; t - \tau, y) - \bar{l}(t, y)\frac{\partial p}{\partial y} + \frac{\overline{l^2}}{2}\frac{\partial^2 p}{\partial y^2},$$

from which follows:

$$\delta\, p(t_0, y_0; t, y) = -\bar{l}(t, y)\frac{\partial p}{\partial y} + \frac{\overline{l^2}}{2}\frac{\partial^2 p}{\partial y^2}. \tag{4}$$

Introduction of Komogoroff's assumptions makes it possible to transform (2) and (4) into the partial differential equations:

$$\frac{\partial p(t_0, y_0; t, y)}{\partial t_0} = -\frac{\partial}{\partial y_0}\left[\frac{\bar{l}}{\tau}(t_0, y_0)\, p\right] - \frac{\partial^2}{\partial y_0^2}\left[\frac{\overline{l^2}}{2\tau} p\right]; \tag{5}$$

$$\frac{\partial p(t_0, y_0; t, y)}{\partial t} = -\frac{\bar{l}}{\tau}(t, y)\frac{\partial p}{\partial y} + \frac{\overline{l^2}}{2\tau}\frac{\partial^2 p}{\partial y^2}. \tag{6}$$

1.4. Equation for the particle-density

We return to the case of variable n, so that $\bar{l}$, $\overline{l^2}$ resp. are defined by 1.21(4), 1.21(5).

By developing the product $n(t-\tau,\, y-l)\;\; P(t-\tau,\, y-l;\, \tau,\, l)$ into a Taylor series and substituting the result into 1.21(4) we obtain:

$$n\,\bar{l} = n\,\overline{l} - \frac{\partial}{\partial y}(n\overline{l^2}). \tag{1}$$

This formula now replaces 1.21(17). Properly speaking the quantities in the right hand member refer to the instant of time $t-\tau$, whereas the left hand member refers to the instant t; it is probable, however, that this difference can be neglected. To the same order of approximation:

$$n\,\overline{l^2} = n\,\overline{l^2}, \tag{2}$$

where the factor n can be dropped, so that this formula is the same as 1.21(18).

A partial differential equation for n can now be derived in two ways. In the first place we can start from 1.1(4), which can be written as follows (with $t_0 = t-\tau$, $y_0 = y-l$):

$$n(t,y) = \int dl\;\, n(t-\tau,\, y-l)\;\; P(t-\tau,\, y-l;\, \tau,\, l). \tag{3}$$

Here the integrand can be developed into the series:

$$n(t-\tau,\, y)\;\; P(t-\tau,\, y;\, \tau,\, l) - l\frac{\partial}{\partial y}(nP) + \frac{l^2}{2}\frac{\partial^2}{\partial y^2}(nP).$$

Hence if from now onward we make at once use of Kolmogoroff's assumptions and pass to the limit $\tau \to 0$:

$$\frac{\partial n}{\partial t} = -\frac{\partial}{\partial y}\left(n\frac{\overline{l}}{\tau}\right) + \frac{\partial^2}{\partial y^2}\left(n\frac{\overline{l^2}}{2\tau}\right). \tag{4}$$

This method of derivation (4) is the analogue of that used before in the derivation of 1.3(6). The equation in this form has already been obtained by Fokker and by Planck; hence the name of equation of Fokker and Planck[1]).

---

[1]) A. D. Fokker, *Ann. d. Physik* (IV) **43**, 812, (1914); and in particular: „Sur les mouvements Browniens dans le champ du rayonnement noir", *Archives Néerlandaises des Sciences exactes, etc.* (IIIA), **4**, 379, (1918).

M. Planck, „Über einen Satz der statistischen Dynamik und seine Erweiterung in der Quantentheorie", *Sitz. Ber. Berliner Akademie*, 324, (1917).

We can start, however, also from the identity which follows from 1.1(3):

$$n(t_0, y) = \int dl\ n(t_0, y)\ P(t_0, y; \tau, l). \tag{5}$$

Here we can develop the integrand into the series:

$$n(t_0, y - l)\ P(t_0, y - l; \tau, l) + l \frac{\partial}{\partial y}(nP) + \frac{l^2}{2} \frac{\partial^2}{\partial y^2}(nP).$$

Substitution of this development into (5), remembering that $t_0 = t - \tau$ and making use of (3), 1.21(4) and 1.21(5), gives:

$$n(t_0, y) = n(t, y) + \frac{\partial}{\partial y}(n\bar{l}) + \frac{\partial^2}{\partial y^2}\left(n \frac{\overline{l^2}}{2}\right),$$

so that at the limit $\tau \to 0$:

$$\frac{\partial n}{\partial t} = -\frac{\partial}{\partial y}\left(n \frac{\bar{l}}{\tau}\right) - \frac{\partial^2}{\partial y^2}\left(n \frac{\overline{l^2}}{2\tau}\right). \tag{6}$$

The equivalence between (4) and (6) is guaranteed by (1) and (2). Equation (6), however, has a formal significance only, as $\bar{l}$ itself is defined through the intermediary of n and $\partial n/\partial y$ according to (1).

It may be remarked that eq. (4) could be read as a relation between the two quantities $\bar{l}$ and $\overline{l^2}$ for the case of variable n; thus replacing eq. 1.21(14) for the isomeric case. Formula (4), however, involves n and its derivatives in a complicated way and as a relation between $\bar{l}$ and $\overline{l^2}$ can have no more than a formal significance. In the case of variable n the phenomenological treatment does not give a direct relation between $\bar{l}$ and $\overline{l^2}$ which could be used to simplify the equations. For constant n (n independent of t as well as of y) eq. (4) of course automatically passes into 1.21(14), while eq. (6) passes into a relation between $\bar{l}$ and $\overline{l^2}$.

Equations (4) and (6) can be written in the form of an equation of continuity:

$$\frac{\partial n}{\partial t} = -\frac{\partial q}{\partial y}, \tag{7}$$

where q, the *intensity of the current*, is defined by:

$$q = n \frac{\bar{l}}{\tau} - \frac{\partial}{\partial y}\left(n \frac{\overline{l^2}}{2\tau}\right). \tag{8}$$

In consequence of eq. (1) above we can write:

$$q = \frac{n}{2}\left(\frac{\overline{l}}{\tau} + \frac{\overline{l}}{\tau}\right), \tag{9}$$

hence it follows that the mean velocity of transportation of the particles can be defined by:

$$\overline{v} = \frac{q}{n} = \frac{1}{2}\left(\frac{\overline{l}}{\tau} + \frac{\overline{l}}{\tau}\right). \tag{10}$$

It thus appears that the diffusion again can be defined by a combination of a flow of dispersion and a flow of concentration.

1.5. Partial differential equation for any property attached to the particles

We will assume that the particles the dispersion of which is controlled by the function p, are carriers of some property w, for example a temperature, or a quantity of suspended materials or the like. It is assumed that every particle in all its displacements takes the value w with it and keeps this value unchanged. (We can extend the considerations if necessary by supposing that the property mentioned is influenced at the same time by exterior agents. Since their action can be expressed by adding certain terms to the equations derived, we will leave aside this case for the moment).

We suppose that the various values of w which are carried by the particles at (t_0, y_0), possess a certain probability distribution, in such a way that $g(w; t_0, y_0)\, dw$ will give the probability that the value of w is to be found between two assigned limits w and $w + dw$. The function g must satisfy the condition

$$\int_{-\infty}^{+\infty} dw\ g(w; t_0, y_0) = 1.$$

The mean value of w for the particles at (t_0, y_0) will then be given by

$$W(t_0, y_0) = \int_{-\infty}^{+\infty} dw\ w\ g(w; t_0, y_0). \tag{1}$$

Since $n(t_0, y_0)\, dy_0$ particles are present in dy_0, we can define the total amount of the property as:

$$W(t_0, y_0)\ n(t_0, y_0)\ dy_0.$$

As the property is carried along by the particles, the function W must satisfy the relation

$$W(t, y)\ n(t, y) = \int_{-\infty}^{+\infty} dy_0\ W(t_0, y_0)\ n(t_0, y_0)\ p(t_0, y_0; t, y). \tag{2}$$

This relation is an extension of form. 1.1(4). It can also be written in the following form:

$$W(t, y)\ n(t, y) = \int_{-\infty}^{+\infty} dl\ W(t_0, y-l)\ n(t_0, y-l)\ P(t_0, y-l; \tau, l). \tag{3}$$

When we apply the same development as was done in the derivation of 1.4(4), the following partial differential equation for W is obtained:

$$\frac{\partial}{\partial t}(nW) = -\frac{\partial}{\partial y}\left(n\overline{\frac{l}{\tau}}\,W\right) + \frac{\partial^2}{\partial y^2}\left(n\overline{\frac{l^2}{2\tau}}\,W\right). \tag{4}$$

Here again we can introduce an intensity of the current Q, defined by:

$$Q = n\overline{\frac{l}{\tau}}\,W - \frac{\partial}{\partial y}\left(n\overline{\frac{l^2}{2\tau}}\,W\right). \tag{5}$$

Equation (4) can be written in the form of an equation of continuity:

$$\frac{\partial}{\partial t}(nW) = -\frac{\partial Q}{\partial y}. \tag{6}$$

1.51. Isomeric diffusion; coefficient of diffusion

Equations 1.5(2); 1.5(4); 1.5(5); 1.5(6) remain valid when we return to the case of constant density. They can be simplified in that case by omitting the constant factor n, so that 1.5(2) and 1.5(4) resp. become:

$$W(t, y) = \int_{-\infty}^{+\infty} dy_0\, W(t_0, y_0)\ p(t_0, y_0; t, y), \tag{1}$$

and

$$\frac{\partial W}{\partial t} = -\frac{\partial}{\partial y}\left(\overline{\frac{l}{\tau}}\,W\right) + \frac{\partial^2}{\partial y^2}\left(\overline{\frac{l^2}{2\tau}}\,W\right). \tag{2}$$

As these equations respectively are of the same form as 1.1(4) or 1.4(3) and 1.4(4), it is seen that the assumption of isomeric dispersion does not lead to a serious restriction of the domain of applicability of the formulae. Equation 1.4(4) indeed can be retained also for isomeric dispersion when it is understood that n does not denote the number of diffusing elements of fluid per unit volume of space, but *e.g.* the number of suspended particles per unit volume of the fluid, *whereas,* $\overline{l}$, $\overline{l^2}$ *etc. refer to the movements of the elements of this fluid,* and thus are connected by eqs. 1.21(16), 1.21(17) and 1.21(18). This is a valuable result, which makes it possible to make use of these relations in many cases. They show that in the case of isomeric dispersion, the degree of turbulence can be completely characterized by one quantity, *viz.* the mean square displacement $\overline{l^2}$. In view of its central importance we shall write:

$$\frac{\overline{l^2}}{2\tau} = \varphi, \tag{3}$$

then:

$$\frac{\overline{l}}{\tau} = \frac{\partial \varphi}{\partial y}; \quad \frac{\overline{l}}{\tau} = -\frac{\partial \varphi}{\partial y}. \tag{4}$$

Introduction of the function φ into eq. (2) transforms it into:

$$\frac{\partial W}{\partial t} = \frac{\partial}{\partial y}\left(\varphi \frac{\partial W}{\partial y}\right). \tag{5}$$

The intensity of the current now assumes the value:

$$Q = -n\varphi \frac{\partial W}{\partial y}. \tag{6}$$

The factor φ evidently plays the part of a „*coefficient of diffusion*".

1.6. Diffusion of particles due to irregular movements combined with a peculiar motion produced by exterior forces

The assumption that the particles, apart from being subjected to irregular motions as considered before, moreover suffer the influence of some exterior force such as for example gravity, which by itself will cause a velocity V, in the greater part of the preceding deductions does not make much difference. In particular with regard to the general case of variable n formulae 1.3(2)—1.3(6), 1.4(1), 1.4(4), 1.4(7)—1.4(10), 1.5(4)—1.5(6) remain valid, as in the derivation of these formulae no suppositions have been introduced concerning either absence or presence of a peculiar motion.

Equations 1.21(14) and 1.51(2) for isomeric dispersion likewise retain their validity. However, the function $f(t_0, t)$ in eq. 1.21(15) now will not be zero, but must take the value V, as in a domain of the field where $\overbracket{l^2}$ has a constant value the only cause which can produce a mean displacement in a definite direction is the peculiar motion. Hence we shall have:

$$\frac{\overbracket{l}}{\tau} = \frac{\partial}{\partial y}\left(\frac{\overbracket{l^2}}{2\tau}\right) + V, \tag{1}$$

for isomeric dispersion. This can be seen also when we eliminate the peculiar motion by introducing a system of coordinates itself moving with the velocity V. Relatively to this system we have irregular motions only with displacements l' related to l by the equation:

$$l' = l - V\tau. \tag{2}$$

Formula 1.21(16) and those following it now remain applicable if we read l' for l, provided τ is sufficiently small in order that we may neglect $\overline{l'}V$ and $V^2\tau$ in comparison with $\overbracket{l'^2}/\tau$. Evidently we have both for variable and for constant n:

$$\frac{\overbracket{l'}}{\tau} = \frac{\overbracket{l}}{\tau} - V; \qquad \frac{\overline{l'}}{\tau} = \frac{\overline{l}}{\tau} - V, \tag{3}$$

$$\frac{\overbracket{l'^2}}{2\tau} = \frac{\overbracket{l^2}}{2\tau}; \qquad \frac{\overline{l'^2}}{2\tau} = \frac{\overline{l^2}}{2\tau}. \tag{4}$$

If for isomeric dispersion we again make use of 1.51(3), we must replace 1.51(4) by

$$\frac{\overbracket{l}}{\tau} = \frac{\partial\varphi}{\partial y} + V; \qquad \frac{\overline{l}}{\tau} = -\frac{\partial\varphi}{\partial y} + V, \tag{5}$$

from which:

$$\frac{1}{2}\left(\frac{\overbracket{l}}{\tau} + \frac{\overline{l}}{\tau}\right) = V. \tag{6}$$

Equation 1.51(2) now can be written:

$$\frac{\partial W}{\partial t} = -\frac{\partial}{\partial y}\left[\left(\frac{\partial\varphi}{\partial y} + V\right)W\right] + \frac{\partial^2}{\partial y^2}(\varphi W), \tag{7}$$

or in slightly different form:

$$\frac{\partial W}{\partial t} = -\frac{\partial}{\partial y}(VW) + \frac{\partial}{\partial y}\left(\varphi\frac{\partial W}{\partial y}\right). \tag{8}$$

The intensity of the current Q then is given by,

$$Q = n\left(\frac{\partial\varphi}{\partial y} + V\right)W - n\frac{\partial}{\partial y}(\varphi W) = n\left(VW - \varphi\frac{\partial W}{\partial y}\right). \tag{9}$$

The fact that in formulae (5), (7) and (9) the peculiar velocity V appears in combination with $\partial\varphi/\partial y$, while evident from the way in which these formula have been obtained, deserves to be noted. We have seen already before that a systematic motion (not due to exterior forces) makes its appearance, when the intensity of the turbulence as represented by $\varphi = \overline{l^2}/2\tau$ is a function of y.

1.7. Introduction of a dispersion function which contains the velocity of the particles

In extension of our former considerations we introduce a new type of dispersion function:

$$\omega(t, y; t + \tau, y + l; v)\, dl\, dv, \tag{1}$$

or in a different notation:

$$\Omega(t, y; \tau, l; v)\, dl\, dv \tag{2}$$

which will denote the probability with which particles situated between y and $y + dy$ at the instant t, will suffer displacements between l and $l + dl$ in the interval from t till $t + \tau$, and at the instant $t + \tau$ will have velocities between v and $v + dv$. This function must satisfy the conditions:

$$\int dv\, \omega(t, y; t + \tau, y + l; v) = p(t, y; t + \tau, y + l), \tag{3}$$

$$\int dv \int dl\ \ \omega(t, y; t + \tau, y + l; v) = 1 \tag{4}$$

where in these and in all following formulae the integrations are extended over the whole domain of available values for v and l respectively.

We define the following mean values:

$$\overline{l} = \int dv \int dl\, l\, \omega(t, y; t + \tau, y + l; v), \tag{5}$$

$$\overline{l^2} = \int dv \int dl\, l^2\, \omega, \tag{6}$$

$$\overline{v} = \int dv \int dl\, v\, \omega, \tag{7}$$

$$\overline{vl} = \int dv \int dl\, v\, l\, \omega. \tag{8}$$

Equations (5) and (6), in which the integration with respect to v

immediately can be carried out, of course are equivalent to 1.21(2) and 1.21(3).

From the relation between the values of v and l for every individual particle and from the fact that the mean values defined in (5)—(8) all refer to the same set of particles, it follows that:

$$\overline{v} = \frac{d\overline{l}}{d\tau}; \quad \overline{vl} = \tfrac{1}{2}\frac{d\overline{l^2}}{d\tau}. \tag{9}$$

Now when K o l m o g o r o f f's assumptions can be applied, we may write

$$\frac{d\overline{l}}{d\tau} = \frac{\overline{l}}{\tau} \quad \text{and} \quad \frac{d\overline{l^2}}{d\tau} = \frac{\overline{l^2}}{\tau}. \tag{10}$$

Hence we obtain:

$$\overline{v} = \frac{\overline{l}}{\tau}; \quad \overline{vl} = \frac{\overline{l^2}}{2\tau}, \tag{11}$$

and consequently for isomeric dispersion:

$$\overline{v} = \frac{d}{dy}\,\overline{vl}. \tag{12}$$

1.71. Expression for the current

The mean velocity $\bar{v}$ of the particles present in the element between y and $y + dy$ at the instant t, for the general case of variable n, is given by the formula:

$$n(t, y)\ \bar{v} = \int dv \int dl\ v\ n(t-\tau, y-l)\ \Omega(t-\tau, y-l; \tau, l; v). \tag{1}$$

When we develop the product $n(t-\tau, y-l)\ \Omega(t-\tau, y-l; \tau, l; v)$ into the series:

$$n(t-\tau, y)\ \Omega(t-\tau, y; \tau, l; v) - l\frac{\partial}{\partial y}(n\Omega) + \frac{l^2}{2}\frac{\partial^2}{\partial y^2}(n\Omega),$$

we obtain:

$$n\,\bar{v} = n\,\overline{v} - \frac{\partial}{\partial y}(n\,\overline{vl}) + \tfrac{1}{2}\frac{\partial^2}{\partial y^2}(n\,\overline{vl^2}).$$

Properly speaking the right hand side of this equation refers to the instant $t-\tau$, but again we take no account of this difference. Further we discard $\overline{vl^2}$, as this must be equal to $\frac{1}{3}(d\,\overline{l^3}/d\tau)$, which will be of the order $\overline{l^3}/3\tau$, and thus according to Kolmogoroff's assumptions is negligible for small values of τ. Finally $n\bar{v}$, from the definition of $\bar{v}$, must be equal to the intensity of the current q. Hence we obtain:

$$q = n\overline{v} - \frac{\partial}{\partial y}(n\,\overline{vl}). \tag{2}$$

Having regard to eqs. 1.7(11) it will be seen that this formula indeed is identical with 1.4(8).

1.72. On the part played by the systematic velocity

It has been remarked in 1.6 that the appearance of a mean displacement of disperson $\overline{l}$ different from zero can be the result of two effects, *viz.* the presence of a ,,*peculiar velocity*'' V due to exterior forces as *e.g.* gravity, and the inhomogenity of the turbulence which makes $\overline{l^2}$ a function of y. In view of 1.7(11) we therefore must conclude that in general $\overline{v}$ will be different from V. This likewise is indicated in the original form of the equation of Fokker and Planck as given by these authors. The point is of importance, as in a paper on the distinction between irregular and systematic motion in diffusion problems Burgers apparently started from the idea that a systematic motion always should be the consequence of the action of exterior forces or of flow in the medium in which the particles are dispersed [1]). Burgers did not make the distinction between ,,systematic'' and ,,peculiar'' motion as introduced at the end of section 1.21, and seems to have assumed that in the absence of exterior forces there would be no systematic motion. This clearly cannot be true under all circumstances, for in that case the first term on the right hand side of 1.71(2) would vanish so that q would reduce to:

$$q = -\frac{\partial}{\partial y}(n\,\overline{vl}).$$

In the case of constant n but variable intensity of turbulence this equation would still lead to a transportation of particles. From the equation of continuity

$$\frac{\partial n}{\partial t} = -\frac{\partial q}{\partial y}$$

---

[1]) J. M. Burgers, ,,On the distinction between irregular and systematic motion in diffusion problems'', *Proc. Acad. Amsterdam*, **44**, 344, (1941).

it would then follow that the density of the particles could vary in the course of time, and a stationary state could be reached only when the product $n\,\overline{vl}$ should be a linear function of y. In the case of a liquid with constant density subjected to the permutation process described in section 1.2 this would lead to a contradiction. This contradiction of course is obviated when we observe that in the case of a liquid of constant density according to eqs. 1.7(11) and 1.6(1) the relation exists:

$$\overline{v} = \frac{\partial}{\partial y}(\overline{vl}) + V, \tag{*}$$

by means of which 1.71(2) is transformed into:

$$q = nV - \overline{vl}\,\frac{\partial n}{\partial y} \tag{**}$$

which properly vanishes when V is zero and n is constant.

We cannot assume general validity for the relations (*) and (**) in all cases. In fact it is known that in the case of a gas in which a non homogeneous field of temperature is kept in existence the condition of equality of pressure necessary for a stationary state requires a non-homogeneous distribution of the density. It consequently cannot be true that the current should depend exclusively upon (*a*) exterior action as evident in V; and (*b*) the gradient of the density $\partial n/\partial y$; in a number of cases the gradient of the turbulence must play an explicit part. We might describe this in general terms by speaking of some kind of ,,*osmotic pressure*'', which depends in part upon the density of the particles, in part upon their state of motion. In the case of the liquid of constant density this ,,osmotic pressure'' comes into play at the slightest tendency towards a deviation of the density from its normal value; in other cases, *e.g.* in a gas, it depends both upon the density and upon the thermal agitation. It must be thought possible therefore that in the case of particles suspended in a liquid of constant density, inequalities in the field of turbulence, while leaving unaffected the density of the liquid itself, nevertheless may produce an unequal distribution of the particle density, in such a way that particles are driven from the regions of strong turbulence and tend to concentrate in regions of weak turbulence.

Whether such an effect will occur in any specific case can be found when the dispersion function $p(t_0, y_0; t, y)$ is given, by calculating the values of $\overline{l}$ and $\overline{l^2}$ and inserting them into eq. 1.4(8). As has been stated already the dispersion function, however, is not more than a purely phenomenological description of what happens, and there is no possibility of distinguishing between the effect of exterior forces and the effect of inequalities of turbulence in producing a non-homogeneous density distribution from a knowledge of this function alone. We must accept it as it stands, combining within itself the result of both effects; it is impossible to calculate the value of V from this function and so to separate out the part played by a peculiar motion, when no other data are available.

Nor does the extended function $\omega(t_0, y_0; t, y; v)$ introduced in section 1.7 promise direct help in this respect. It can be used in order to calculate the values of $\overline{v}$ and $\overline{vl}$, but as is seen from 1.7(9) these values do not give a new feature, as they can be obtained just as well from the function $p(t_0, y_0; t, y)$. The function ω may be used also to calculate quantities like $\overline{v^2}$; to decide, however, whether such quantities will reveal to us features of the dispersion process which could be used to make a distinction between effects of inhomogeneous turbulence and peculiar motion, will be possible only when a physical theory is available, which gives us a real insight into what is happening and not a mere phenomenological description of the results of this process.

CHAPTER 2

EXAMPLES OF DISPERSION FUNCTIONS ILLUSTRATING CERTAIN PROPERTIES DEDUCED IN CHAPTER I

2.1. An elementary example of isomeric dispersion

The differential equation 1.4(4) for the particle-density and the relation 1.21(16) which must be fulfilled by the quantities $\overline{l}$ and $\overline{l^2}$ in order that this equation should admit the solution n = *constant*, can be illustrated by constructing some simple examples of diffusion processes.

An elementary case is obtained as follows: let us take, at $t = 0$, a system of particles, lying with a constant density n_0 on the positive side of the plane $y = 0$. To the particles lying between y and $y + dy$ we first give displacements:

$$l = \pm \beta y, \tag{1}$$

where β is a constant smaller than 1, one half of the particles being displaced in the positive direction, the other half in the negative direction. We suppose the dimensions of the particles to be so small in comparison with their distances, that no attention need be given to collisions.

In this case we shall have:

$$\overline{l} = 0, \quad \overline{l^2} = \beta^2 y^2, \tag{2}$$

hence the relation 1.21(16) is *not* fulfilled. Equation 1.4(4) taken in its original form

$$\delta n = -\frac{\partial}{\partial y}(n_0 \overline{l}) + \frac{\partial^2}{\partial y^2}\left(n_0 \frac{\overline{l^2}}{2}\right) {}^{1)} \tag{3}$$

gives:

$$\delta n = \frac{\partial^2}{\partial y^2}\left(n_0 \frac{\beta^2 y^2}{2}\right) = n_0 \beta^2. \tag{4}$$

---

[1]) In this form the validity of the integral relation 1.11(3) is not involved, as no assumption has been used concerning a time dependence of the dispersion.

The increase in density thus calculated, can be verified directly. In consequence of the displacements (1) for half of the particles the density is decreased to

$$\frac{n_0}{1+\beta},$$

while for the other half it is increased to

$$\frac{n_0}{1-\beta}.$$

Hence the average density of the system becomes

$$\frac{n_0}{2}\left(\frac{1}{1+\beta}+\frac{1}{1-\beta}\right)=\frac{n_0}{1-\beta^2},$$

so that:

$$\delta n = \frac{n_0}{1-\beta^2} - n_0 = \frac{n_0\beta^2}{1-\beta^2} \backsimeq n_0\beta^2 \qquad (5)$$

if β is sufficiently small.

It will be evident how we can construct a case in which the density remains constant, *viz.* by giving to all particles an additional displacement

$$l_a = +\beta^2 y,$$

by which an expansion is produced which brings back the density to n_0. The resulting displacements of the particles now become:

$$l = (\beta + \beta^2)\, y \quad \text{and} \quad l = (-\beta + \beta^2)\, y, \qquad (6)$$

with equal chances for + and — signs. In this case:

$$\overline{l} = \beta^2\, y; \quad \overline{l^2} \backsimeq \beta^2\, v^2. \qquad (7)$$

In the expression for $\overline{l^2}$ a term $\beta^4 y^2$ again has been neglected in comparison with $\beta^2 y^2$; it appears that to this order of approximation (which corresponds to that observed in Chapter 1) the relation 1.21(16) between $\overline{l}$ and $\overline{l^2}$ is fulfilled.

Equation (3) now reduces to:

$$\delta \iota = -\frac{\partial}{\partial y}(n_0\beta^2 y) + \frac{\partial^2}{\partial y^2}\left(n_0\frac{\beta^2 y^2}{2}\right),$$

which for constant n_0 gives

$$\delta n = 0,$$

so that the dispersion described by eqs. (6) in fact is isomeric.

The fact that the introduction of the additional displacement $\beta^2 y$ in this case is an arbitrary procedure, illustrates the purely phenomenological character of the dispersion function, which can be constructed for any case we choose.

2.2. Generalization

We can generalize the example of the preceding section by starting from a dispersion function of the type:

$$p_0 = \frac{\alpha}{\sqrt{\pi}\,\psi(y_0)}\, e^{-\frac{\alpha^2(y-y_0)^2}{\psi(y_0)^2}}. \tag{1}$$

This function fulfils the relation

$$\int_{-\infty}^{+\infty} dy\, p_0 = 1 \tag{2}$$

and gives:

$$\overline{l} = \overline{y - y_0} = 0; \quad \overline{l^2} = \overline{(y-y_0)^2} = \frac{\psi(y_0)^2}{2\alpha^2}. \tag{3}$$

We suppose α to be a large quantity (of the order $\tau^{-\frac{1}{2}}$); the function ψ on the contrary will be of normal magnitude.

From eq. 2.1(3), again starting with a constant density n_0, we deduce:

$$\delta n = n_0 \frac{\partial^2}{\partial y^2}\left(\frac{\psi^2}{4\alpha^2}\right). \tag{4}$$

The change of the density can be eliminated by giving to each element of volume dy an expansion in the ratio

$$1 + \frac{\partial^2}{\partial y^2}\left(\frac{\psi^2}{4\alpha^2}\right),$$

which leads to an extra displacement:

$$l_a = \int \frac{\partial^2}{\partial y^2}\left(\frac{\psi^2}{4\alpha^2}\right) dy = \frac{\partial}{\partial y}\left(\frac{\psi^2}{4\alpha^2}\right) = \frac{\psi\psi'}{2\alpha^2}, \tag{5}$$

which is of the order of magnitude α^{-2} (or τ). Writing for the sake of brevity:

$$\chi = \frac{\psi\psi'}{2\alpha^2}; \tag{5a}$$

we next consider the dispersion function:

$$p = \frac{\alpha}{\sqrt{\pi}\,\psi(y_0)}\, e^{-\frac{\alpha^2\{y-y_0-\chi(y_0)\}^2}{\psi(y_0)^2}}. \tag{6}$$

This function again fulfils:

$$\int_{-\infty}^{+\infty} dy\, p = 1, \tag{7}$$

while now:

$$\overline{l} = \overline{y - y_0} = \chi; \quad \overline{l^2} = \frac{\psi^2}{2\alpha^2} + \chi^2 \cong \frac{\psi^2}{2\alpha^2}, \tag{8}$$

as χ^2 (which is of the order α^{-4} or τ^2) may be neglected in comparison with $\psi^2/2\alpha^2$.

If the above reasoning is correct, then the function p defined by (6) should fulfil the relation:

$$\int_{-\infty}^{+\infty} dy_0\, p = 1. \tag{9}$$

In verifying this relation we take note of the circumstance that the domain of values of y_0 which give a substantial contribution to the integral, will be small, as $y_0 - y$ at most will be of the order α^{-1}. Consequently we shall make use of series developments in which we restrict to terms of the order α^{-2} (which also is the order of χ). In the first place we have:

$$\alpha^2\{y - y_0 - \chi(y_0)\}^2 = \alpha^2\{\chi(y) + (y_0 - y)\,(1 + \chi') + \tfrac{1}{2}(y_0 - y)^2 \chi''\}^2,$$

where here and in what follows the functions χ, ψ, χ' etc. when written without any argument will refer to the argument y. The expression between { } can be simplified by introducing a new variable ε so that:

$$y_0 - y = \frac{(\varepsilon - \chi)}{(1 + \chi')}.$$

As ε can be treated as being of the order α^{-1}, the expression considered reduces to:

$$\alpha^2\,\{\varepsilon + \tfrac{1}{2}\varepsilon^2\chi''\}^2 \cong \alpha^2\varepsilon^2,$$

the terms discarded being at least of the order α^{-3}. The exponent can then be written:

$$-\frac{\alpha^2\varepsilon^2}{\{\psi(y) + (y_0 - y)\,\psi' + \tfrac{1}{2}(y_0 - y)^2\,\psi''\}^2} =$$

$$= -\frac{\alpha^2\varepsilon^2}{\psi^2} + \frac{2\alpha^2\varepsilon^2(\varepsilon - \chi)\,\psi'}{\psi^3} - \frac{\alpha^2\varepsilon^4}{\psi^4}\,(3\psi'^2 - \psi\psi'').$$

Leaving the first term apart the rest of the exponent is at most of the order α^{-1}; hence the exponential function corresponding to it can be developed into:

$$1+\frac{2\alpha^2\varepsilon^2(\varepsilon-\chi)\,\psi'}{\psi^3}+\frac{2\alpha^4\varepsilon^6\psi'^2}{\psi^6}-\frac{\alpha^2\varepsilon^4}{\psi^4}\,(3\psi'^2-\psi\psi'').$$

We further develop the factor $1/\psi(y_0)$ occurring in the integrand before the exponential functions as follows:

$$\frac{1}{\psi(y_0)}=\frac{1}{\psi}\left\{1-\frac{(\varepsilon-\chi)\,\psi'}{\psi}+\frac{\varepsilon^2}{\psi^2}\,(\psi'^2-\tfrac{1}{2}\psi\psi'')\right\}.$$

The integral finally takes the form:

$$\frac{\alpha}{\sqrt{\pi}\,(1+\chi')\,\psi}\int\limits_{-\infty}^{+\infty}d\varepsilon\,e^{-\frac{\alpha^2\varepsilon^2}{\psi^2}}\left\{1-\frac{(\varepsilon-\chi)\psi'}{\psi}+\frac{2\alpha^2\varepsilon^2(\varepsilon-\chi)\psi'}{\psi^3}-\right.$$

$$\left.-\frac{2\alpha^2\varepsilon^4\psi'^2}{\psi^4}+\frac{\varepsilon^2}{\psi^2}\,(\psi'^2-\tfrac{1}{2}\psi\psi'')+\frac{2\alpha^4\varepsilon^6\psi'^2}{\psi^6}-\frac{\alpha^2\varepsilon^4}{\psi^4}\,(3\psi'^2-\psi\psi'')\right\}.$$

Working out the integration we obtain the result:

$$1-\chi'+\frac{\psi'^2}{2\alpha^2}+\frac{\psi\psi''}{2\alpha^2}.$$

This will be equal to 1 provided:

$$\chi'=\frac{\psi'^2+\psi\psi''}{2\alpha^2},\tag{10}$$

which is equivalent to (5*a*) above [1]).

2.3. Dispersion function fulfilling the integral condition 1.11(3)

In the construction of the dispersion functions mentioned in the preceding sections, no attention has been given to the condition of

---

[1]) An isomeric dispersion function can also be obtained by means of the following formula:

$$p=\frac{d\eta}{dy}\sqrt{\frac{k}{\pi}}\,e^{-k(\eta-y_0)^2}$$

where k is a function of y_0, while η is a function of y defined by:

$$\frac{dy}{d\eta}=\int\limits_{-\infty}^{+\infty}dy_0\sqrt{\frac{k}{\pi}}\,e^{-k(\eta-y_0)^2}.$$

integrability which was brought forward in 1.11(3). A function satisfying this condition is given *e.g.* by the formula:

$$p(t_0, y_0; t, y) = \frac{d\eta}{dy}\sqrt{\frac{k}{\pi\tau}}\, e^{-\frac{k(\eta-\eta_0)^2}{\tau}} \tag{1}$$

where k is a constant; $\tau = t - t_0$; $\eta = f(t, y)$, *i.e.* an arbitrary function of t and y; while $\eta_0 = f(t_0, y_0)$. This function p at the same time fulfils the condition:

$$\int_{-\infty}^{+\infty} p\, dy = 1. \tag{2}$$

Integration with respect to dy_0, however, in general will not give the value unity, and hence the dispersion process described by this function will not be isomeric. By way of example we take:

$$\eta = y^{1/3};\quad \eta_0 = \eta_0^{1/3}, \tag{3}$$

so that

$$p = \frac{1}{3y^{2/3}}\sqrt{\frac{k}{\pi\tau}}\, e^{-\frac{k(y^{1/3}-y_0^{1/3})^2}{\tau}}. \tag{4}$$

This function will be applied over the whole domain from $y = -\infty$ to $y = +\infty$, assuming that $\eta = y^{1/3}$ is negative when v is negative, whereas $y^{2/3}$ always will be positive. The singularity of p for $y = 0$ will be considered as unimportant, as p remains integrable; it probably can be obviated by taking a more complicated function for y, but this would make all calculations rather cumbersome.

Integrations to be performed on the function p with respect to y can be carried out most conveniently by observing that:

$$p\, dy = 3p\eta^2\, d\eta = \sqrt{\frac{k}{\pi\tau}}\, e^{-\frac{k(\eta-\eta_0)^2}{\tau}}\, d\eta. \tag{5}$$

As:

$$l = y - y_0 = (\eta - \eta_0)^3 + 3(\eta - \eta_0)^2\,\eta_0 + 3(\eta - \eta_0)\,\eta_0^2,$$

we find:

$$\overline{l} = \int_{-\infty}^{+\infty} dl\ l\ p = \frac{3}{2}\frac{\eta_0\tau}{k}, \tag{6}$$

$$\overline{l^2} = \int_{-\infty}^{+\infty} dl\ l^2\ p = \frac{15}{8}\frac{\tau^3}{k^3} + \frac{45}{4}\frac{\eta_0^2\tau^2}{k^2} + \frac{9}{2}\frac{\eta_0^4\tau}{k}. \tag{7}$$

Hence for $\tau \to 0$, writing y for y_0, we obtain:

$$\lim \frac{\overline{l}}{\tau} = \frac{3}{2} \frac{y^{1/3}}{k}, \tag{8}$$

$$\lim \frac{\overline{l^2}}{2\tau} = \frac{9}{4} \frac{y^{4/3}}{k}. \tag{9}$$

Direct calculation shows that with these values the partial differential equations 1.3(5) and 1.3(6) for the function p are fulfilled.

The relation 1.21(16) evidently is not fulfilled. The expression for the current becomes:

$$q = \frac{3}{2} \frac{y^{1/3}}{k} n - \frac{\partial}{\partial y}\left(\frac{9}{4} \frac{y^{4/3}}{k} n\right). \tag{10}$$

A stationary density distribution can be obtained when q is independent of y. The solution of eq. (10) for this case is:

$$n = C_1 y^{-1/3} + C_2 y^{-2/3}, \tag{11}$$

with:

$$q = -\frac{3}{4} \frac{C_1}{k}. \tag{12}$$

The expression (7) for $\overline{l^2}$ shows that for large values of the time interval the mean square displacement, in this example, increases proportionally with τ^3.

2.4. Dispersion function fulfilling the three conditions:

$$\int dy \; p = 1, \tag{1}$$

$$\int dy_0 \; p = 1, \tag{2}$$

$$p(t_0, y_0; t, y) = \int dz \; p(t_0, y_0; t'', z) \; p(t'', z; t, y) \tag{3}$$

where $t_0 < t'' < t$. When a dispersion function fulfils these conditions simultaneously, the mean displacements $\overline{l}$, $\bar{l}$ and the mean square displacement $\overline{l^2}$ derived from it must fulfil the relations:

$$\frac{\bar{l}}{\tau} = -\frac{\overline{l}}{\tau}; \tag{4}$$

$$\frac{\overline{l}}{\tau} = \frac{\partial}{\partial y}\left(\frac{\overline{l^2}}{2\tau}\right). \tag{5}$$

Because of the importance of the isomeric dispersion process for the study of incompressible fluids it may be worth while to find an example of a dispersion function which fulfils the conditions and properties above, and which allows us to study, in a more concrete way, the properties of isomeric dispersion functions and of mean values in connection with the results obtained in Chapter 1.

The simplest case of a dispersion function fulfilling equations (1) (3) is G a u s s' function:

$$p = \frac{1}{\sqrt{\pi b \tau}} e^{-\frac{(y-y_0)^2}{b\tau}} \tag{6}$$

where b is a constant. However, as, in consequence of its symmetrical character, it gives:

$$\overline{l} = 0, \quad \overline{l} = 0 \quad \text{and} \quad \frac{\overline{l^2}}{2} = \text{constant}, \tag{7}$$

it presents a rather trivial case. All dispersion functions which have a symmetrical character are useless for our purpose. After a number of preliminary attempts it was finally found possible, however, by introducing a generalised form of (6), of unsymmetrical character, to obtain a satisfactory result. As it is useful to be also able to verify the relations of section 1.7, we shall, at once, start with the ω-function, from which the p-function is obtained by means of formula 1.7(3). For this ω-function we shall take the expression:

$$\omega \, dv \, dy = \sqrt{\frac{k}{\pi^2 \tau}} \frac{d\eta}{dy} (1 + \beta_0 \tau + \gamma_0 \lambda + \delta_0 \lambda^2) \, e^{-\frac{\lambda^2}{\tau} - k(v-\theta)^2} \, dv \, dy, \tag{8}$$

the notations having the following meanings:

η = monotonous function of y, varying just as y from $-\infty$ to $+\infty$, with finite derivatives and $d\eta/dy$ different from zero for all values of y,

$\tau = (t - t_0)$. (numerical factor); it is assumed to be a very small quantity,

$\lambda = \eta - \eta_0$; it is practically restricted to a region having a width of the order of $\sqrt{\tau}$,

$\beta_0, \gamma_0, \delta_0, \theta$ = functions of y_0, to be determined from the conditions which must be fulfilled by $p = \int_{-\infty}^{+\infty} dv \, \omega$ and by ω itself; it will be found that they are given by the following expressions:

$$\gamma_0 = -\frac{\eta_0''}{2\eta_0'^2}, \quad \delta_0 = -2\beta_0 = \frac{\gamma_0^2}{2} + \frac{\gamma_0'}{2\eta_0'},$$

$$\theta = \frac{\lambda}{2\eta_0'\tau} - \frac{\eta_0''}{8\eta_0'^3} - \frac{\eta_0''\lambda^2}{2\eta_0'^3\tau},$$

k = a very small constant, at most of the order of the minimum value to be allowed for τ.

By carrying out the integration with respect to v we obtain the dispersion function for the displacements alone, as follows:

$$p = \int_{-\infty}^{+\infty} dv\, \omega = \frac{1}{\sqrt{\pi\tau}} \frac{d\eta}{dy} (1 + \beta_0\tau + \gamma_0\lambda + \delta_0\lambda^2)\, e^{-\frac{\lambda^2}{\tau}}. \qquad (9)$$

We shall investigate the relations (1)—(5) with the aid of this formula.

2.41. Relations fulfilled by the function p

I. For constant y_0 (and therefore also constant η_0, β_0, γ_0, δ_0):

$$\int_{-\infty}^{+\infty} dy\, p = \int_{-\infty}^{+\infty} d\lambda \frac{1}{\sqrt{\pi\tau}} (1 + \beta_0\tau + \gamma_0\lambda + \delta_0\lambda^2)\, e^{-\frac{\lambda^2}{\tau}} =$$

$$= 1 + \beta_0\tau + \tfrac{1}{2}\delta_0\tau.$$

In order that this expression shall be equal to 1, it is necessary that

$$\beta_0 = -\tfrac{1}{2}\delta_0. \qquad (1)$$

II. We calculate $\overline{l}$ and $\overline{l^2}$ by the following formulae:

$$\overline{l} = \int_{-\infty}^{+\infty} dy\, (y - y_0)\, p, \quad \overline{l^2} = \int_{-\infty}^{+\infty} dy\, (y - y_0)^2\, p.$$

Here y can be considered as a function of λ, which can be developed into the series:

$$y = y_0 + \frac{\lambda}{\eta_0'} - \frac{\lambda^2}{2} \frac{\eta_0''}{\eta_0'^3} - \frac{\lambda^3}{6} \left(\frac{\eta_0'''}{\eta_0'^4} - \frac{3\eta_0''^2}{\eta_0'^5} \right). \qquad (2)$$

With the approximation that terms with higher powers than λ^2 are negligible we find:

$$\frac{\overline{l}}{\tau} = \frac{\gamma_0}{2\eta_0'} - \frac{\eta_0''}{4\eta_0'^3}; \quad \frac{\overline{l^2}}{2\tau} = \frac{1}{4\eta_0'^2}.$$

The relation

$$\frac{\overline{l}}{\tau} = \frac{\partial}{\partial y_0} \left(\frac{\overline{l^2}}{2\tau} \right)$$

consequently requires:

$$\gamma_0 = -\frac{\eta_0''}{2\eta_0'^2}. \qquad (3)$$

Hence:

$$\frac{\overline{l}}{\tau} = -\frac{\eta_0''}{2\eta_0'^3}; \quad \frac{\overline{l^2}}{2\tau} = \frac{1}{4\eta_0'^2}: \qquad (4)$$

III. In the calculation of $\int_{-\infty}^{+\infty} dy_0\, p$, y must be considered as a constant. Now y_0 can be considered as a function of λ, which can be developed into the following series:

$$y_0 = y - \frac{\lambda}{\eta'} - \frac{\lambda^2}{2}\frac{\eta''}{\eta'^3} + \frac{\lambda^3}{6}\left(\frac{\eta'''}{\eta'^4} - \frac{3\eta''^2}{\eta'^5}\right).$$

Here η', η'', η''' refer to the constant value of y and consequently are likewise constants. We also require corresponding developments of β_0, γ_0, δ_0. However, in the factor $\{1 + \beta_0\tau + \gamma_0\lambda + \delta_0\lambda^2\}$ it is sufficient to develop γ_0 only; we have:

$$\gamma_0 = \gamma + \gamma'(y_0 - y) = \gamma - \frac{\gamma'\lambda}{\eta'},$$

so that the factor becomes:

$$\left\{1 + \beta\tau + \gamma\lambda + \left(\delta - \frac{\gamma'}{\eta'}\right)\lambda^2\right\}.$$

We now have to calculate:

$$\int_{-\infty}^{+\infty} d\lambda \left\{1 + \lambda\frac{\eta''}{\eta'^2} - \frac{\lambda^2}{2}\left(\frac{\eta'''}{\eta'^3} - \frac{3\eta''^2}{\eta'^4}\right)\right\}\left\{1 + \beta\tau + \gamma\lambda + \left(\delta - \frac{\gamma'}{\eta'}\right)\gamma^2\right\} \cdot \frac{e^{-\frac{\lambda^2}{\tau}}}{\sqrt{\pi\tau}}.$$

After simple integrations of G a u s s' functions, the expression reduces to unity, so that, indeed:

$$\int_{-\infty}^{+\infty} dy_0\, p = 1. \tag{5}$$

IV. For isomeric dispersion we have:

$$\bar{l} = \int_{-\infty}^{+\infty} dy_0\,(y - y_0)\, p =$$

$$= \int_{-\infty}^{+\infty} d\lambda \left\{1 + \lambda\frac{\eta''}{\eta'^2} - \ldots\right\}\left(\frac{\lambda}{\eta'} + \frac{\lambda^2}{2}\frac{\eta''}{\eta'^3}\right)(1 + \gamma\lambda)\frac{e^{-\frac{\lambda^2}{\tau}}}{\sqrt{\pi\tau}} = \frac{1}{2}\frac{\eta''\tau}{\eta'^3}, \tag{6}$$

so that:

$$\frac{\bar{l}}{\tau} = -\frac{\overline{l}}{\tau} = \frac{\eta''}{2\eta'^3}. \tag{7}$$

V. We finally calculate the integral of the product in 2.4(3):

$$\int_{-\infty}^{+\infty} dy_1 \frac{d\eta_1}{dy_1}(1 + \beta_0\tau_1 + \gamma_0\lambda_1 + \delta_0\lambda_1^2)\frac{e^{-\frac{\lambda_1^2}{\tau_1}}}{\sqrt{\pi\tau_1}} \cdot$$

$$\cdot \frac{d\eta}{dy}(1 + \beta_1\tau_2 + \gamma_1\lambda_2 + \delta_1\lambda_2^2)\frac{e^{-\frac{\lambda_2^2}{\tau_2}}}{\sqrt{\pi\tau_2}},$$

where

$$\lambda_1 = \eta_1 - \eta_0, \quad \lambda_2 = \eta - \eta_1.$$

When we put:

$$\tau = \tau_1 + \tau_2, \quad \eta_1 = \mu + \frac{(\eta\tau_1 + \eta_0\tau_2)}{\tau},$$

we have:

$$\lambda_1 = \mu + (\eta - \eta_0)\frac{\tau_1}{\tau} = \mu + \nu_1,$$

$$\lambda_2 = -\mu + (\eta - \eta_0)\frac{\tau_2}{\tau} = -\mu + \nu_2,$$

and the exponent in the e-function becomes:

$$-\mu^2\frac{\tau}{\tau_1\tau_2} - \frac{(\eta - \eta_0)^2}{\tau}.$$

Further β_0, γ_0, δ_0 are constant, while β_1, γ_1, δ_1 are variable; from the latter γ_1 only has to be developed:

$$\gamma_1 = \gamma_0 + (\gamma_1 - \gamma_0)\gamma_0' = \gamma_0 + \lambda_1\frac{\gamma_0'}{\eta_0'}.$$

After substitution for λ_1, λ_2 we must calculate:

$$\frac{d\eta}{dy}\frac{1}{\pi\sqrt{\tau_1\tau_2}}\int_{-\infty}^{+\infty} d\mu \Big\{1 + \beta_0\tau + \gamma_0(\eta - \eta_0) +$$

$$+\left(\frac{\gamma_0'}{\eta_0'} + \gamma_0^2\right)(-\mu^2 - \mu\nu_1 + \mu\nu_2 + \nu_1\nu_2) +$$

$$+\delta_0(2\mu^2 + 2\mu\nu_1 - 2\mu\nu_2 + \nu_1^2 + \nu_2^2)\Big\} \cdot e^{-\mu^2\frac{\tau}{\tau_1\tau_2} - \frac{(\eta-\eta_0)^2}{\tau}} =$$

$$= \frac{d\eta}{dy}\frac{1}{\sqrt{\pi\tau}}e^{-\frac{(\eta-\eta_0)^2}{\tau}}\Big\{1 + \beta_0\tau + \gamma_0(\eta - \eta_0) +$$

$$+\left(\gamma_0^2 + \frac{\gamma_0'}{\eta_0'}\right)\nu_1\nu_2 + \delta_0(\nu_1^2 + \nu_2^2) - \left(\gamma_0^2 + \frac{\gamma_0'}{\eta_0'} - 2\delta_0\right)\frac{\tau_1\tau_2}{2\tau}\Big\}.$$

In order that the right hand member shall be equal to p, it is necessary that the expression between $\{\ldots.\}$ take the value:

$$1 + \beta_0\tau + \gamma_0(\eta - \eta_0) + \delta_0(\eta - \eta_0)^2.$$

This will be the case if:

$$\delta = \frac{\gamma^2}{2} + \frac{\gamma'}{2\eta'}, \tag{8}$$

so that:

$$\delta_0 = -\frac{\eta_0'''}{4\eta_0'^3} + \frac{5}{8}\frac{\eta_0''^2}{\eta_0'^4}. \tag{9}$$

We observe that with the value (8) of δ, we obtain:

$$(1 + \beta_0\tau_1 + \gamma_0\lambda_1 + \delta_0\lambda_1^2)(1 + \beta_1\tau_2 + \gamma_1\lambda_2 + \delta_1\lambda_2^2) =$$

$$= 1 + \beta_0\tau + \gamma_0(\eta - \eta_0) + \delta_0(\eta - \eta_0)^2. \tag{10}$$

Hence the dispersion function 2.4(9) fulfils all the conditions and properties required.

2.42. Integro-differential equation fulfilled by the function $\omega(t_0, y_0; t, y; v)$

We now return to the function ω, which involves the velocities of the particles at the instant t. In order to find the value to be assigned to the quantity θ, we shall first deduce an integro-differential equation to be fulfilled by ω in order to guarantee the proper connection between v and l. We consider a large number N of particles which at the instant t_0 have started from the region between y_0 and $y_0 + dy_0$. The values of the position coordinate y and of the velocity v which these particles have acquired at the instant t can be represented by means of points in a diagram, having y and v as coordinates. The number of points in any element $dy\, dv$ of this diagram will be equal to $N\,\omega(t_0, y_0; t, y, v)\, dy\, dv$.

The particles which at the instant t have their representative points in an element $dy\, dv$, at the earlier instant $t_1 = t - dt$ will have had their representative points in other elements $dy_1\, dv_1$ of the same diagram, in such a way that the difference of position coordinates $y - y_1$ can be taken equal to $v_1 dt$, where v_1 is the velocity of the particle at the instant $t - dt$. It is not possible to give a relation between the velocity v_1 at this instant and the velocity v at the instant t, as the velocity in general will have suffered irregular changes in the interval dt, in no way correlated with the values of v_1 or y_1. It is assumed, however, that these changes are of the order of dt, or at least that their mean value is of the order of dt, so that the expression $v_1 dt$ is a sufficient approximation to the value of $y - y_1$. Hence all the particles which at the instant t have their representative points in the element $dydv$, at the earlier instant will have had their representative points distributed over a series of elements dy_1dv_1, determined by the equation: $y_1 = y - v_1\, dt$. We shall call this series of elements dy_1dv_1 the series II.

When we consider another element $dy\, dv$ of the y, v-diagram for the instant t, having *the same value of* y as the element $dy\, dv$ considered first, and ask for the positions of the representative points of the particles at the earlier instant $t - dt$, we shall find that these representative points were situated in the same series II of elements $dy_1\, dv_1$. It follows that the total number of particles

$$N\, dy \int_{-\infty}^{+\infty} dv\, \omega(t_0, y_0; t, y, v),$$

which at the instant t have their representative points in a series of elements $dy\, dv$ all belonging to one definite value of y (but corresponding to all the various possible values of v), which series we shall indicate as the series I, at the earlier instant $t - dt$ will have their representative points in elements of the series II.

On the other hand when we consider any one of the elements dy_1dv_1 of the series II, we shall find that the particles which, at the instant $t - dt$, had their representative points in such an element, at the later instant t will have their representative points moved to elements $dy\, dv$ for which $y = y_1 + v_1\, dt$, while the values of v may have changed in an irregular way. That is to say, the representative points of these particles have moved to elements of the series I.

From this we conclude that all the particles which at the instant $t - dt$ had their representative points in elements of the series II, will have their representative points in elements of the series I at the later instant t; while, on the contrary, all particles which have their representative points in elements $dy\, dv$ of the series I at the instant t, must have come from the elements of the series II. It follows that the total number of representative points in both series of elements must be the same. As evidently the number N plays no part in these considerations, and $dy_1 = dy$ for constant v, we have the relation:

$$\int_{-\infty}^{+\infty} dv\, \omega(t_0, y_0; t, y, v) = \int_{-\infty}^{+\infty} dv_1\, \omega(t_0, y_0; t - dt, y - v_1\, dt; v_1), \qquad (1)$$

from which

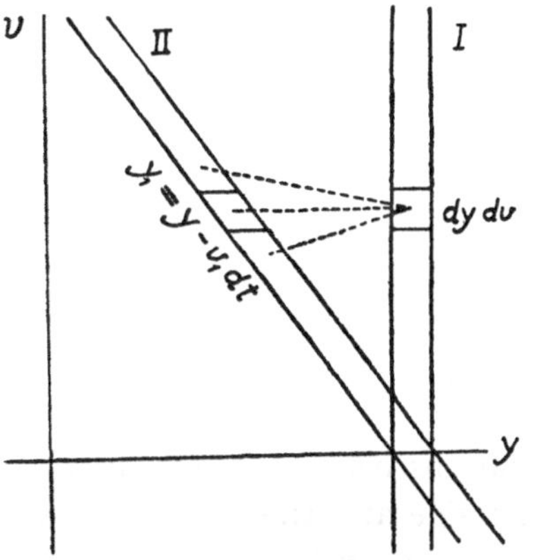

Fig. 2.42.

$$\int_{-\infty}^{+\infty} dv \left(\frac{\partial \omega}{\partial t} + v \frac{\partial \omega}{\partial y} \right) = 0. \quad (2)$$

Making use of the relation

$$\int_{-\infty}^{+\infty} dv\, v\, \omega = p\, \theta \quad (3)$$

we obtain:

$$\frac{\partial p}{\partial t} + \frac{\partial}{\partial y} (p\, \theta) = 0. \quad (4)$$

This equation can be used to calculate θ. After substitution for p, this equation gives the following approximate expression for θ, keeping to terms of the orders $\frac{1}{\sqrt{\tau}}$ and 1 (detailed calculations will be omitted):

$$\theta = \frac{\lambda}{2\eta_0' \tau} - \frac{\eta_0''}{8\eta_0'^3} - \frac{\eta_0'' \lambda^2}{2\eta_0'^3 \tau}. \quad (5)$$

2.43. Some mean values involving the velocities

By putting $\tau = t - t_0$, $l = y - y_0$, $p(t_0, y_0; t, y)$ can be written as $p(t_0, y_0; t_0 + \tau, y_0 + l)$, and eq. 2.42(4) becomes:

$$\frac{\partial p}{\partial \tau} + \frac{\partial}{\partial l} (p\, \theta) = 0.$$

When we multiply both terms in the left hand member of this equation by l and integrate with respect to l, we find:

$$\int_{-\infty}^{+\infty} dl\, l \frac{\partial p}{d\tau} = \frac{d}{d\tau} \int_{-\infty}^{+\infty} dl\, l\, p = \frac{d\overline{l}}{d\tau};$$

$$\int_{-\infty}^{+\infty} dl\, l \frac{\partial}{\partial l} (p\theta) = l\, p\, \theta \Big|_{-\infty}^{+\infty} - \int_{-\infty}^{+\infty} dl\, p\, \theta.$$

The first term of the right hand member is zero, as p vanishes for $l = \pm \infty$. The second term, according to 2.42(3), is equal to:

$$- \int_{-\infty}^{+\infty} dl \int_{-\infty}^{+\infty} dv\, v\, \omega$$

and thus, in consequence of 1.7(7) has the value: $-\overline{v}$.

Hence it follows that:
$$\frac{d\overline{l}}{d\tau} = \overline{v}. \tag{1}$$

In an analogous way we obtain:

$$\int_{-\infty}^{+\infty} dl\, l^2 \frac{\partial p}{\partial t} = \frac{d\overline{l^2}}{d\tau},$$

$$\int_{-\infty}^{+\infty} dl\, l^2 \frac{\partial}{\partial l}(p\,\theta) = -2\,\overline{vl},$$

so that

$$\frac{d\overline{l^2}}{d\tau} = 2\,\overline{vl}. \tag{2}$$

Equations (1) and (2) are the relations between the mean values involving the displacements on one side and the mean values involving the velocities on the other. They can be checked easily by calculating explicitly the values of the integrals 1.7(7) and 1.7(8); we find as results (detailed calculations being omitted):

$$\overline{v} = \int_{-\infty}^{+\infty} dl\, p\,\theta = -\frac{\eta_0''}{2\eta_0'^3}, \tag{3}$$

$$\overline{vl} = \int_{-\infty}^{+\infty} dl\, p\,\theta\, l = \frac{1}{4\eta_0'^2}\,. \tag{4}$$

By comparing these values with those of $\overline{l}/\tau$ and $\overline{l^2}/2\tau$ found in 2.41(4) we see that relations (1) and (2) are verified.

In 1.71 we have already defined the resulting velocity of transportation by:

$$\overline{v} = \int_{-\infty}^{+\infty} dy \int_{-\infty}^{+\infty} dv\, v\, p'(t_0, y_0; t, y, v).$$

In the present case of isomeric dispersion, where $p' = p$, we find:

$$\overline{v} = \int_{-\infty}^{+\infty} dy_0 \int_{-\infty}^{+\infty} dv\, v \sqrt{\frac{k}{\pi^2\tau}}\, \frac{d\eta}{dy}\,(1 + \beta_0\tau + \gamma_0\lambda + \delta_0\lambda^2)\, e^{-\frac{\lambda^2}{\tau} - k(v-\theta)^2},$$

here β_0, γ_0, δ_0 and θ all vary with y_0. Integration with respect to v gives:

$$\overline{v} = \int_{-\infty}^{+\infty} dy_0 \frac{1}{\sqrt{\pi\tau}}\, \frac{d\eta}{dy}\,(1 + \beta_0\tau + \gamma_0\lambda + \delta_0\lambda^2)\, e^{-\frac{\lambda^2}{\tau}}\,\theta =$$

$$= \int_{-\infty}^{+\infty} d\lambda\, \frac{e^{-\frac{\lambda^2}{\tau}}}{\sqrt{\pi\tau}} \left(1 + \lambda\frac{\eta''}{\eta'^2} - \ldots\right)(1 + \beta\tau + \gamma\lambda + \ldots)\,\cdot$$

$$\cdot \left(\frac{\lambda}{2\eta'\tau} + \frac{\lambda^2\eta''}{2\eta'^3\tau} - \frac{\eta''}{8\eta'^3} - \frac{\eta''\lambda^2}{2\eta'^3\tau}\right) \text{ (comp. p. 41, III).}$$

After effectuation of the various Gaussian integrals we obtain:

$$\bar{v} = 0. \tag{5}$$

This is correct because in an isomeric dispersion, there is no diffusion flow $q = n\bar{v}$. The result is in accordance with the formula 1.4(10):

$$\bar{v} = \tfrac{1}{2}\left(\frac{\overline{l}}{\tau} + \frac{\bar{l}}{\tau}\right),$$

because: $\bar{l} = -\overline{l}$.

Finally:

$$\overline{v^2} = \int_{-\infty}^{+\infty} dl \int_{-\infty}^{+\infty} dv\, v^2\, \omega = \int_{-\infty}^{+\infty} dl \left(\frac{1}{2k} + \theta^2\right) p = \frac{1}{2k} + \frac{1}{8\eta'^2\tau}. \tag{6}$$

This formula shows that $\overline{v^2}$ increases when τ decreases. However, we cannot decrease τ indefinitely to zero, as there is a finite interval of correlation. When k is very small in comparison with the value of τ, then $\overline{v^2}$ will be practically constant.

2.44. Investigation of the product integral for the function ω

Let us write:

$$\omega = \sqrt{\frac{k}{\pi^2\tau}}\,\eta'(1 + \beta_0\tau + \gamma_0\lambda + \delta_0\lambda^2)\, e^{-\frac{\lambda^2}{\tau} - k(v-\theta)^2};$$

$$\omega_I = \sqrt{\frac{k}{\pi^2\tau_1}}\,\eta_1'(1 + \beta_0\tau_1 + \gamma_0\lambda_1 + \delta_0\lambda_1^2)\, e^{-\frac{\lambda_1^2}{\tau_1} - k(v_1-\theta_0)^2};$$

$$\omega_{II} = \sqrt{\frac{k}{\pi^2\tau_2}}\,\eta'(1 + \beta_1\tau_2 + \gamma_1\lambda_2 + \delta_1\lambda_2^2)\, e^{-\frac{\lambda_2^2}{\tau_2} - k(v-\theta_1)^2};$$

with:

$$\lambda = \eta - \eta_0, \quad \lambda_1 = \eta_1 - \eta_0, \quad \lambda_2 = \eta - \eta_1;$$

$$\beta_0, \gamma_0, \delta_0 = \text{functions of } \eta_0 \text{ or } y_0;$$

$$\beta_1, \gamma_1, \delta_1 = \text{functions of } \eta_1 \text{ or } y_1;$$

$$\theta_0 = \frac{\lambda_1}{2\eta_0'\tau_1} - \frac{\eta_0''}{8\eta_0'^3} - \frac{\eta_0''\lambda^2}{2\eta_0'^3\tau_1}; \quad \theta_1 = \frac{\lambda_2}{2\eta_1'\tau_2} - \frac{\eta_1''}{8\eta_1'^3} - \frac{\eta_1''\lambda^2}{2\eta_1'^3\tau_2}.$$

It is asked to prove that the integral

$$\int_{-\infty}^{+\infty} dy_1 \int_{-\infty}^{+\infty} dv_1\, \omega_I\, \omega_{II} = \omega \tag{1}$$

for constant values of y_0, y and v.

By the integration with respect to v_1, the factor

$$\sqrt{\frac{k}{\pi}}\, e^{-k(v_1-\theta_0)^2}$$

disappears from the product $\omega_I \omega_{II}$. In the integration with respect to y_1 we must take into account that terms of higher orders than τ can be neglected; therefore we can write:

$$\beta_1 = \beta_0;$$

$$\gamma_1 = \gamma_0 + \gamma_0'(y_1 - y) = \gamma_0 + \gamma_0' \frac{\lambda_1}{\eta_0'};$$

$$\delta_1 = \delta_0;$$

$$\theta_1 = \frac{\lambda_2}{2\eta_0'\tau_2} - \frac{\lambda_1\lambda_2\eta''}{2\eta_0'^3\tau_2} - \frac{\eta_0''}{8\eta_0'^3} - \frac{\eta_0''\lambda_2^2}{2\eta_0'^3\tau_2}.$$

We shall omit the detailed calculations and mention only that the following transformation of variables is introduced:

$$\tau_1 + \tau_2 = \tau, \quad \lambda_1 = \mu + \nu_1, \quad \lambda_2 = -\mu + \nu_2$$

where:

$$\nu_1 = (\eta - \eta_0)\frac{\tau_1}{\tau}, \quad \nu_2 = (\eta - \eta_0)\frac{\tau_2}{\tau}, \quad \nu_1 + \nu_2 = \eta - \eta_0.$$

Then according to 2.41(10):

$$(1 + \beta_0\tau_1 + \gamma_0\lambda_1 + \delta_0\lambda_1^2)\,(1 + \beta_1\tau_2 + \gamma_1\lambda_2 + \delta_1\lambda_2^2) =$$

$$= 1 + \beta_0\tau + \gamma_0(\eta - \eta_0) + \delta_0(\eta - \eta_0)^2.$$

This factor is thus a constant and has the form needed in the resulting function ω. The exponent becomes:

$$-\left(\frac{\lambda_1^2}{\tau_1} + \frac{\lambda_2^2}{\tau_2}\right) - k(v - \theta_1)^2 =$$

$$= -\mu^2\frac{\tau}{\tau_1\tau_2} - k(v + \alpha\mu - \theta)^2 - \frac{(\eta - \eta_0)^2}{\tau}$$

where:

$$\alpha = \frac{1}{2\eta_0'\tau_2} - \frac{(\eta - \eta_0)\eta_0''}{2\eta_0'^3\tau_2} = \text{order}\,\frac{1}{\tau};$$

$$\theta = \frac{(\eta - \eta_0)}{2\eta_0'\tau} - \frac{\eta_0''}{8\eta_0'^3} - \frac{\eta_0''(\eta - \eta_0)^2}{2\eta_0'^3\tau}.$$

The left hand member of (1) consequently takes the following form:

$$\sqrt{\frac{k}{\pi^3\tau_1\tau_2}}\,(1+\beta_0\tau+\gamma_0\lambda+\delta_0\lambda^2)\,e^{-\frac{\lambda^2}{\tau}}\int_{-\infty}^{+\infty} d\mu\, e^{-\frac{\mu^2\tau}{\tau_1\tau_2}-k(v+\alpha\mu-\theta)^2} =$$

$$=\sqrt{\frac{k}{\pi^2\tau}}\,(1+\beta_0\tau+\gamma_0\lambda+\delta_0\lambda^2)\,e^{-\frac{\lambda^2}{\tau}}\,.\,\frac{e^{-\frac{k(v-\theta)^2}{1+k^*}}}{\sqrt{1+k^*}},$$

where

$$k^* = k\,\frac{\alpha^2\tau_1\tau_2}{\tau}\left(=\text{order}\,\frac{k}{\tau}\right).$$

In the end result we see that there is still an influence of τ_1 and τ_2 in the distribution of velocities. However, integration with respect to v gives the right result for p, as terms depending on τ_1, τ_2 will then disappear. Further if k has been taken very small with respect to τ, the influence of τ_1, τ_2 will be very much reduced and becomes practically unimportant.

2.5. Isomeric dispersion function constructed from Gauss' error function by introducing a discontinuous change in b

It has been attempted to construct a dispersion function by solving the partial differential equations:

$$\frac{\partial p}{\partial t}=\frac{\partial}{\partial y}\left[\varphi(t,y)\,\frac{\partial p}{\partial y}\right], \tag{1}$$

$$\frac{\partial p}{\partial t_0}=-\frac{\partial}{\partial y_0}\left[\varphi(t_0,y_0)\,\frac{\partial p}{\partial y_0}\right], \tag{2}$$

where φ is a given function, for which the following course has been taken:

$$\begin{aligned}&\text{for } y<0: \quad \varphi=\frac{a}{4}\\ &\text{for } y>0: \quad \varphi=\frac{b}{4},\end{aligned} \tag{3}$$

a and b being constants. The expression for p obtained in this way has different forms, depending upon the signs of y and y_0. We therefore write:

$$
y_0 < 0 \begin{cases} y < 0: \; p_1 = \dfrac{1}{\sqrt{\pi a \tau}} e^{-\frac{(y-y_0)^2}{a\tau}} + \dfrac{A}{\sqrt{\pi a \tau}} e^{-\frac{(y+y_0)^2}{a\tau}} & (4) \\[2ex] y > 0: \; p_2 = \dfrac{B}{\sqrt{\pi b \tau}} e^{-\frac{(y-\alpha)^2}{b\tau}} & (5) \end{cases}
$$

$$
y_0 > 0 \begin{cases} y < 0: \; p_1^* = \dfrac{B^*}{\sqrt{\pi a \tau}} e^{-\frac{(y-\alpha^*)^2}{a\tau}} & (6) \\[2ex] y > 0: \; p_2^* = \dfrac{1}{\sqrt{\pi b \tau}} e^{-\frac{(y-y_0)^2}{b\tau}} + \dfrac{A^*}{\sqrt{\pi b \tau}} e^{-\frac{(y+y_0)^2}{b\tau}} & (7) \end{cases}
$$

In order to determine the values to be given to the coefficients A, B, A^*, B^*, α, α^*, condition 2.4(1) furnishes two equations (one for $y_0 < 0$ and one for $y_0 > 0$), while the circumstance that both the p-function itself and the function $\varphi\, \partial p/\partial y$ must be continuous at $y = 0$ furnishes 4 other equations:

$$
p_1 = p_2; \quad p_1^* = p_2^*,
$$

$$
\left(\varphi \frac{\partial p}{\partial y}\right)_1 = \left(\varphi \frac{\partial p}{\partial y}\right)_2; \quad \left(\varphi \frac{\partial p^*}{\partial y}\right)_1 = \left(\varphi \frac{\partial p^*}{\partial y}\right)_2.
$$

These 6 equations determine the 6 coefficients; we find:

$$
\left.\begin{array}{l}
\left.\begin{array}{l} A = \dfrac{\sqrt{a} - \sqrt{b}}{\sqrt{a} + \sqrt{b}} \\[2ex] B = \dfrac{2\sqrt{b}}{\sqrt{a} + \sqrt{b}} \end{array}\right\} \quad A + B = 1. \\[6ex]
\left.\begin{array}{l} A^* = -\dfrac{\sqrt{a} - \sqrt{b}}{\sqrt{a} + \sqrt{b}} \\[2ex] B^* = \dfrac{2\sqrt{a}}{\sqrt{a} + \sqrt{b}} \end{array}\right\} \quad A^* + B^* = 1. \\[6ex]
\alpha = k y_0 \quad \text{where } k = \sqrt{\dfrac{b}{a}} \quad (y_0 < 0) \\[3ex]
\alpha^* = k^* y_0 \quad \text{where } k^* = \sqrt{\dfrac{a}{b}} \quad (y_0 > 0).
\end{array}\right\} \qquad (8)
$$

For the function so defined we have investigated whether the relations 2.4(1)—2.4(5) are fulfilled. This was easily found to be the case for 2.4(1)—2.4(3).

More attention was necessary in the case of 2.4(4) and 2.4(5). It was found that $\overline{l}/\tau$, $\overline{l}/\tau$ and $\partial/\partial y(\overline{l^2}/2\tau)$ all become zero for $\tau=0$, so that formally eqs. 2.4(4) and 2.4(5) appear to be fulfilled. Equation 2.4(4) is moreover fulfilled for small non zero values of τ. This is not the case, however, with eq. 2.4(5). This result was somewhat disappointing; it may be connected with the discontinuity in φ. Perhaps it must be considered as a warning that Kolmogoroff's assumptions are still somewhat schematic. As the integrations already in the present case become very cumbersome, the calculation will not reproduced here; for the same reason we have refrained from seeking other forms for p.

Chapter 3

TIME MEAN VALUES CONNECTED WITH THE MOTION OF A PARTICLE

3.1. Introductory remarks

We may distinguish between two methods for investigating the irregular motion of particles. The first one is to consider a great number of particles simultaneously and to introduce a probability function for the distribution of the displacements or for that of the velocities. In this method the behaviour of every individual particle is not considered. This is the statistical method applied in Chapter 1.

In the second method the history of a single particle will be studied. For this purpose the velocity v of the particle must be considered as a more or less irregular function of the time t, and the mean value will be calculated over a certain interval of time for this particular particle. It is the object of the present chapter to investigate mean values of this kind. An important problem will then be to find a connection between the two methods.

Along with the function $v(t)$ itself we have to consider functions of v, say $G(v)$, obtained by performing certain operations upon v. We must distinguish between functions containing only v and its derivatives, and functions containing also integrals of v, as for instance the displacement of the particle. The necessity for this distinction is to be found in the circumstance that provided a certain central assumption is fulfilled, any non fractional function $G(v)$ containing the velocity or derivatives of the velocities always remains finite, whereas a function containing the displacement may increase indefinitely with the time.

3.11. Central assumption concerning the behaviour of the function $v(t)$

Consider a more or less irregularly oscillating function of the time, $v(t)$. It will be assumed that this function presents a stationary character in the statistical sense. This assumption will form the basis of all deductions given in this chapter. Though its general meaning is not difficult to be seen, it contains so much that it is not easy to express it by means of any single formula. It leads to an infinity of consequences, apparently independent from each other, every one of which illustrates an aspect of the stationary statistical character of $v(t)$.

A very general category of these aspects can be described as follows: Let $G(v)$ be an arbitrary function of v and its derivatives $(\dot{v}, \ddot{v}, \ldots)$, but not containing any integral of v. At every instant t the function G must have a finite and definite value. We then calculate the quantity:

$$\bar{G}(t_0, T) = \frac{1}{T} \int_{t_0}^{t_0+T} dt_1 \, G(t_1). \tag{1}$$

to be called the ***mean value*** of G. The assumption concerning the stationary character of $v(t)$ in the statistical sense then requires that the quantity $\bar{G}$, with sufficient approximation, can be considered as constant, independent of both t_0 and T, provided T surpasses a certain limit T'. The words „with sufficient approximation" mean that the variations of $\bar{G}$ with respect to t_0 and T must remain below certain limits, sufficiently small (in view of the purpose of the calculations) in comparison either with $\bar{G}$ or with the amplitude of the oscillations of $G(t)$.

The simplest cases are obtained by putting G respectively equal to v^2, $\dot{v}^2$, $\ddot{v}^2$, ... Then we must require *e.g.* that the variations of the quantity:

$$\overline{v^2} = \frac{1}{T} \int_{t_0}^{t_0+T} dt_1 \, \{v(t_1)\}^2 \tag{2}$$

are sufficiently small in comparison with $\overline{v^2}$ itself. An analogous condition can be given with respect to $\dot{v}^2$, $\ddot{v}^2$, ...

It can be expected that when the condition is fulfilled with $G =$ resp. v^2, $\dot{v}^2$, $\ddot{v}^2$, ..., it automatically will also be fulfilled for $G =$ resp. $|v|$, $|\dot{v}|$, $|\ddot{v}|$, ... and for $G = v, \dot{v}, \ddot{v}$, ...; likewise for $G = \dot{v}v$, etc.

The class of functions of v which are enclosed in the form of G considered above does not contain all functions without integrals of v which we have to introduce in statistical investigations. We have also to consider functions of the type:

$$v(t)\, v(t+\eta),$$

and we must require that the mean value of such a function likewise will be independent of t_0 and T, provided $T > T'$. This, however, will be a consequence of the assumptions already made, as is easy to prove by a development of $v(t+\eta)$ into a series proceeding according to the increasing powers of η.

3.12. Correlations

Consider the mean value:

$$S(\eta) = \overline{v(t)\, v(t+\eta)} = \frac{1}{T}\int_{t_0}^{t_0+T'} dt_1\, v(t_1)\, v(t_1+\eta). \qquad (1)$$

This function will be called the *correlation of the function $v(t)$ for the interval η*. Evidently when η is zero, we shall have:

$$S(o) = \overline{v^2}. \qquad (2)$$

We introduce the *coefficient of correlation*, $R(\eta)$, by means of the formula:

$$R(\eta) = \frac{\overline{v(t)\, v(t+\eta)}}{\overline{v(t)^2}} = \frac{S(\eta)}{S(o)}. \qquad (3)$$

It can easily be proved that $S(\eta)$ [and similarly, $R(\eta)$] is an even function of η. Indeed, as $S(\eta)$ is independent of t_0, we have the identity:

$$S(\eta) = \frac{1}{T}\int_{t_0}^{t_0+T'} dt_1\, v(t_1)\, v(t_1+\eta) = \frac{1}{T}\int_{t_0+\eta}^{t_0+\eta+T'} dt_1\, v(t_1)\, v(t_1+\eta) =$$

$$= \frac{1}{T}\int_{t_0}^{t_0+T'} dt_2\, v(t_2-\eta)\, v(t_2) = S(-\eta). \qquad (4)$$

3.121. Functions with limited duration of correlation

An important class of functions $v(t)$ is of such nature that the correlation $S(\eta)$ has a value appreciably differing from zero only for a limited domain of values of the time interval η, so that we can write:

$$S(\eta) = 0 \text{ for } \eta > \vartheta. \qquad (1)$$

In this case the new quantity ϑ, which is smaller than T' (usually T' must be chosen much larger than ϑ), is the so-called *upper measure of correlation* to be found in the course of $v(t)$.

The degree of approximation of eq. (1) must be such that

$$\int_0^\infty d\eta\, S(\eta) \quad \text{is convergent.} \tag{2}$$

In many cases it is possible to define ϑ in such a way that it can be taken as upper limit of the integral in (2) without impairing its value.

In order that (1) may be valid we must have:

$$\int_{t_0}^{t_0+T} dt_1\, v(t_1)\, v(t_1+\eta) < m \qquad \text{for } \eta > \vartheta \tag{3}$$

where m is a finite quantity such that m/T' is negligible in comparison with $\overline{v^2}$; this inequality must apply for arbitrary values of t_0 and T.

Introducing the coefficient of correlation $R(\eta)$ we shall write:

$$\Theta = \int_0^\infty d\eta\, R(\eta) \cong \int_0^\vartheta d\eta\, R(\eta). \tag{4}$$

This quantity has the dimension of a time and will be called the *average duration of correlation*. As $R(\eta) \leqslant 1$, we have $\Theta < \vartheta$.

It is useful also to introduce the correlation integral:

$$\widehat{vv} = \int_0^\infty d\eta\, \overline{v(t)\, v(t+\eta)} = \overline{v^2}\, \Theta. \tag{5}$$

3.122. Corollary

Consider the integral:

$$I(\eta) = \frac{1}{T} \int_{t_0+\eta}^{t_0+T} dt_1\, v(t_1)\, v(t_1 \pm \eta), \tag{1}$$

where it is assumed that $T > T'$ as before. The integral will reduce to the type already considered, when the denominator before the integral is changed into $T - \eta$. We therefore write:

$$I(\eta) = \frac{T-\eta}{T} \frac{1}{T-\eta} \int_{t_0+\eta}^{t_0+T} dt_1\, v(t_1)\, v(t_1+\eta) = \frac{T-\eta}{T} S(\eta). \tag{2}$$

This formula will hold as long as $T - \eta$ is sufficiently large (larger than T'). Now $S(\eta)$ will differ from zero only when $\eta < \vartheta$, in which case $\eta \ll T$. In that case we can replace $(T - \eta)/T$ by unity, so that:

$$I(\eta) = S(\eta) = \overline{v^2}\, R(\eta). \tag{3}$$

For large values of η the functions $S(\eta)$ and $R(\eta)$ vanish. This will also be the case with $I(\eta)$, as can be seen from formula (1) by observing that the integral itself will be smaller than m when η surpasses ϑ, and that m/T can be neglected in comparison to $\overline{v^2}$. Hence formula (2) and even formula (3) may be used generally for all values of η.

3.123. Proof that $\bar{v} = 0$ for functions fulfilling 3.121(1)

It will be evident that 3.121(1) and 3.121(3) can be valid only when $v(t)$ changes sign sufficiently often in such a way that neither positive nor negative values will be preponderant. It can be proved indeed that the mean value of $\bar{v}$ must be zero in this case. We have:

$$\bar{v} = \frac{1}{T}\int_{t_0}^{t_0+T} dt_1\, v(t_1),$$

and consequently

$$(\bar{v})^2 = \left\{\frac{1}{T}\int_{t_0}^{t_0+T} dt_1\, v(t_1)\right\}^2 = \frac{2}{T^2}\int_{t_0}^{t_0+T} dt_2 \int_{t_2}^{t_0+T} dt_1\, v(t_1)\, v(t_2).$$

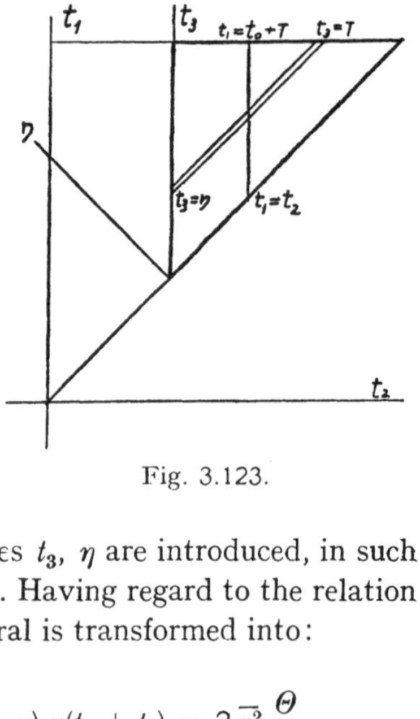

Fig. 3.123.

In the double integral new variables t_3, η are introduced, in such a way that: $\eta = t_1 - t_2$; $t_3 = t_1 - t_0$. Having regard to the relation $\partial(\eta, t_3)/\partial(t_1, t_2) = 1$, the double integral is transformed into:

$$(\bar{v})^2 = \frac{2}{T^2}\int_0^T d\eta \int_\eta^T dt_3\, v(t_0 + t_3 - \eta)\, v(t_0 + t_3) = 2\,\overline{v^2}\,\frac{\Theta}{T}.$$

As $\overline{v^2}$ and Θ are finite quantities, independent of T, $(\bar{v})^2$ can be made as small as we require by increasing T. Hence we conclude:

$$(\bar{v})^2 = 0;\ \bar{v} = 0. \tag{1}$$

3.13. Correlation when $\bar{v}$ is different from zero

In the case where $\bar{v}$ is not equal to zero, we write:

$$v'(t) = v(t) - \bar{v}, \tag{1}$$

The quantity v' may be called the *fluctuation* of v. Formula (1) implies that

$$\overline{v'} = 0. \tag{2}$$

Instead of the quantity $S(\eta)$ introduced in 3.12(1) we define a new quantity:

$$\sigma(\eta) = \overline{v'(t)\, v'(t+\eta)} = \frac{1}{T} \int_{t_0}^{t_0+T} dt_1\, v'(t_1)\, v'(t_1+\eta). \tag{3}$$

The coefficient of correlation then will be given by:

$$\varrho(\eta) = \frac{\overline{v'(t)\, v'(t+\eta)}}{\overline{v'(t)^2}} = \frac{\sigma(\eta)}{\sigma(o)}. \tag{4}$$

The new functions $\sigma(\eta)$ and $\varrho(\eta)$ again will be even functions of η; further the measure of correlation can be defined by the condition

$$\sigma(\eta) = 0 \quad \text{for } \eta > \vartheta, \tag{5}$$

or by:

$$\left| \int_{t_0}^{t_0+T} dt_1\, v'(t_1)\, v'(t_1+\eta) \right| < m' \text{ for } \eta > \vartheta, \tag{6}$$

while the average duration of correlation will be given by:

$$\theta = \int_0^\infty d\eta\, \varrho(\eta). \tag{7}$$

In consequence of the assumption that $\bar{v}$ is a constant, the following relations exist between the functions $S(\eta)$ and $\sigma(\eta)$, resp. $R(\eta)$ and $\varrho(\eta)$, which again are a consequence of the assumption that $\bar{v}$ is a constant:

$$\overline{v^2} = \overline{v'^2} + (\bar{v})^2 \tag{8}$$

$$S(\eta) = \sigma(\eta) + (\bar{v})^2 = \overline{v'^2}\, \varrho(\eta) + (\bar{v})^2 \tag{9}$$

$$R(\eta) = \frac{\overline{v'^2}}{\overline{v^2}} \varrho(\eta) + \frac{(\bar{v})^2}{\overline{v^2}} = 1 - \frac{\overline{v'^2}}{\overline{v^2}} (1 - \varrho). \tag{10}$$

The function Θ defined in 3.121 (4) with the upper limit ∞ is divergent, but for any finite value of t exceeding ϑ we shall have:

$$\Theta_t = \int_0^t d\eta\, R(\eta) = \frac{\overline{v'^2}}{\overline{v^2}}\,\theta + \frac{(\bar{v})^2}{\overline{v^2}}\,t, \tag{11}$$

and likewise:

$$\widehat{vv}_t = \int_0^t d\eta\, \overline{v(t)\, v(t+\eta)} = \overline{v'^2}\,\theta + (\bar{v})^2\, t. \tag{12}$$

Finally we obtain the following extension of the corollary given in section 3.122:

$$I(\eta) = \frac{1}{T}\int_{t_0+\eta}^{t_0+T} dt_1\, v(t_1)\, v(t' \pm \eta) = \frac{1}{T}\int_{t_0+\eta}^{t_0+T} dt_1\, v'(t_1)\, v'(t_1 \pm \eta) +$$

$$+ \frac{1}{T}\int_{t_0+\eta}^{t_0+T} dt_1\, (\bar{v})^2 = \frac{T-\eta}{T}\,\sigma(\eta) + \frac{T-\eta}{T}\,(\bar{v})^2. \tag{13}$$

In the first term, which differs from zero only for $\eta < \vartheta$, we may replace $(T-\eta)/T$ by unity. Hence:

$$I(\eta) = \sigma(\eta) + \frac{T-\eta}{T}\,(\bar{v})^2 = S(\eta) - \frac{\eta}{T}\,(\bar{v})^2 =$$

$$= \overline{v^2}\,\varrho(\eta) + \frac{T-\eta}{T}\,(\bar{v})^2 = \overline{v^2}\,R(\eta) - \frac{\eta}{T}\,(\bar{v})^2. \tag{14}$$

To shorten the notation we write:

$$\frac{\overline{v'^2}}{\overline{v^2}} = \psi, \quad \frac{(\bar{v})^2}{\overline{v^2}} = 1 - \psi. \tag{15}$$

Then:

$$R(\eta) = \psi\,\varrho(\eta) + (1-\psi) \tag{16}$$

$$\Theta_t = \psi\,\theta + (1-\psi)\,t. \tag{17}$$

Important particular cases of these formulae are obtained:

(*a*) when the fluctuations v' are very small in comparison to the mean value $\bar{v}$, in which case ψ will be a very small quantity;

(*b*) when the mean value $\bar{v}$ is small in comparison to the amplitude of the fluctuations, in which case ψ will be nearly equal to unity.

3.14. Relative magnitudes of intervals of time

In the preceding sections we have introduced two characteristic intervals of time:

T' (in 3.11), being a lower limit for the duration over which time mean values are calculated according to form. 3.11(1);

ϑ (in 3.121), representing an upper measure for the correlation in the course of the function $v(t)$.

The appropriate values to be given to T' and ϑ are dependent upon the nature of $v(t)$. In view of the part played by T in our formulae the quantity T' must be chosen considerably larger than ϑ. In cases where the function $v(t)$ presents a stationary character in the statistical sense, as has been assumed throughout the preceding deductions, the value of T in form. 3.11(1) can be taken as large as we choose, so that there is no upper limit to T. This, however, will not be the case when $v(t)$ is not stationary in the statistical sense. In that case mean values taken over a finite duration $T > T'$ with various values of t_0 generally will differ amongst themselves, and the application of the preceding analysis becomes much more difficult. In certain cases it may be possible to define the value of T' in such a way that minor fortuitous variations (variations of short period) are eliminated from the mean value $\overline{G}$ in a satisfactory way, while at the same time it can serve as an upper limit which must not be exceeded by T in order that important changes of $\overline{G}$ extending over longer periods are not lost sight of.

T can be materially restricted to the order of ϑ ($T \geq \vartheta$), by considering, according to a method to be introduced in 3.3, a great number of particles starting simultaneously at the instant t_0 from the same element of volume, and then taking the average of the mean values referring to the individual particles.

3.2. Mean values in which the displacement plays a part

We assume the function $v(t)$ to represent the v locity of a particle and will investigate a number of relations in which the position of the particle is involved.

When the position of the particle at any instant t is represented by $y(t)$, we have: $dy/dt = v(t)$, and inversely:

$$y(t) = y(t_0) + \int_{t_0}^{t} dt_1\, v(t_1). \tag{1}$$

It is useful to introduce also the displacement:

$$l(t_0, t) = \int_{t_0}^{t} dt_1\, v(t_1) = y(t) - y(t_0). \tag{2}$$

From the expressions (1) and (2) for the position and the displacement we can calculate the following mean values, which frequently occur in statistical investigations:

$$\bar{l},\ \bar{y};\ \overline{l^2},\ \overline{y^2};\ \overline{vl},\ \overline{vy}.$$

3.21. Investigation of $\bar{l}$ and $\bar{y}$

For a first orientation it is useful to begin with the particular case v = constant = a; we then have:

$$l = a(t - t_0)$$

$$\bar{l} = \frac{aT}{2}, \tag{1}$$

so that $\bar{l}$ increases indefinitely with T. The condition that $v(t)$ will have a stationary statistical character is clearly not sufficient to ensure the existence of a proper mean value for $\bar{l}$.

It can be surmised that this likewise will be the case when v itself is not constant, but has a constant value $\bar{v}$ differing from zero.

In fact by taking the mean value of l:

$$\bar{l}(t_0; T) = \frac{1}{T}\int_{t_0}^{t_0+T} dt \int_{t_0}^{t} dt_1\, v(t_1),$$

and differentiating $\bar{l}$ with respect to T we obtain:

$$\frac{\partial \bar{l}}{\partial T} = -\frac{\bar{l}}{T} + \frac{1}{T}\int_{t_0}^{t_0+T} dt_1\, v(t_1) = -\frac{\bar{l}}{T} + \bar{v}, \tag{2}$$

which equation is valid provided $T > T'$. It may be considered as a differential equation for $\bar{l}$, and we can obtain its integral in the form:

$$\bar{l} = \frac{\bar{v}}{2} T + \frac{C}{T}, \tag{3}$$

where C is the integration constant, provided $v(t)$ is stationary in the statistical sense, so that $\bar{v}$ is constant. Hence for sufficiently large values of T we may generally assume:

$$\frac{\bar{l}}{T} = \frac{\bar{v}}{2}. \tag{4}$$

As: $y = l + y(t_0)$, and consequently

$$\bar{y} = \bar{l} + y(t_0),$$

we have:

$$\frac{\partial \bar{y}}{\partial T} = \frac{\bar{v}}{2}. \tag{5}$$

From this it can be seen that the validity of the central assumption is not sufficient to ensure the existence of a constant mean value $\bar{y}$; such a mean value is to be obtained only when the further condition $\bar{v} = 0$ is fulfilled.

This result reveals an instance of the difference in behaviour between functions depending exclusively upon v and its derivatives, and functions containing also integrals of v. It is, however, not the only difference, for whereas the condition $\bar{v} = 0$ makes possible the existence of a constant mean value $\bar{l}$, it will be seen in the next section that even then $\overline{l^2}$ appears to increase linearly with T.

3.22. Investigation of $\overline{l^2}$

Equation 3.2(2) defining l can be written:

$$l = \int_0^{t_2} d\eta \, v(t_0 + \eta), \tag{1}$$

where $t_2 = t - t_0$; hence the mean square displacement is given by:

$$\overline{l^2} = \frac{1}{T}\int_0^T dt_2 \int_0^{t_2} d\eta_1 \int_0^{t_2} d\eta_2 \, v(t_0 + \eta_1)\, v(t_0 + \eta_2) =$$

$$= \frac{1}{T}\int_0^T dt_2 \int_0^{t_2} d\eta_1 \int_0^{t_2} d\eta_2 \, v(t_0 + t_2 - \eta_1)\, v(t_0 + t_2 - \eta_2). \tag{2}$$

The domain of integration is given by the quadrilateral pyramid

$O.ABCD$ depicted in fig. 3.22, or, in consequence of the symmetry between η_1 and η_2, by twice the triangular pyramid $O.ABC$. After introduction of new variables:

$$t = t_2,\ \eta = \eta_1,\ \delta = \eta_2 - \eta_1,$$

which make:

$$\frac{\partial(t, \eta, \delta)}{\partial(t_2, \eta_1, \eta_2)} = 1,$$

the triple integral changes into:

$$\overline{l^2} = 2\int_0^T d\delta \int_\delta^T d\eta \int_\eta^T dt\, \frac{v(t_0 + t - \eta)\, v(t_0 + t - \eta + \delta)}{T} \quad . \tag{3}$$

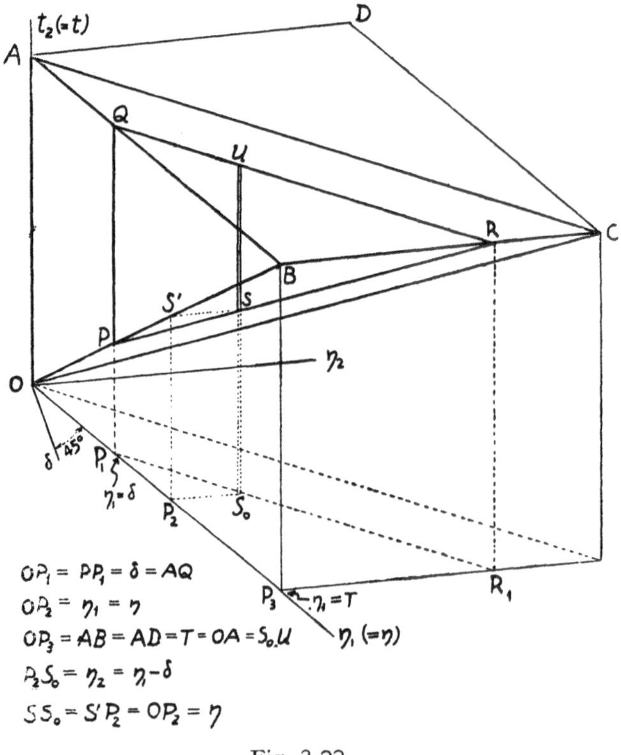

Fig. 3.22.

The integral with respect to t may be put equal to

$$\frac{T - \eta}{T}\, \overline{v^2}\, R(\delta).$$

This certainly is valid if $T - \eta > T'$. The formula, however, can also be applied if η approaches T, because then both the integral and the factor $(T - \eta)/T$ tend to zero.

Further integration and division by T now gives:

$$\frac{\overline{l^2}}{T} = 2\int_0^T d\delta \int_\delta^T d\eta \frac{T-\eta}{T^2} \overline{v^2} R(\delta) = \overline{v^2}\int_0^T d\delta \left(1 - \frac{2\delta}{T} + \frac{\delta^2}{T^2}\right) R(\delta). \quad (4)$$

Now with the notation introduced in 3.13 we write:

$$R(\delta) = \psi\, \varrho(\delta) + (1 - \psi),$$

where $\varrho(\delta) = 0$ for $\delta > \vartheta$. When T is much larger than ϑ, effectuation of the integration in (4) gives:

$$\frac{\overline{l^2}}{T} = \psi\, \overline{v^2} \int_0^T d\delta\, \varrho(\delta) + (1 - \psi)\, \overline{v^2} \int_0^T d\delta \left(1 - \frac{2\delta}{T} + \frac{\delta^2}{T^2}\right) =$$

$$= \psi\, \overline{v^2}\theta + \tfrac{1}{3}(1 - \psi)\, \overline{v^2} T. \quad (5)$$

This can also be written:

$$\frac{\overline{l^2}}{T} = \overline{v'^2}\theta + \tfrac{1}{3}(\bar{v})^2\, T. \quad (6)$$

In the case $\bar{v} = 0$ this reduces to:

$$\frac{\overline{l^2}}{T} = \overline{v^2}\,\theta. \quad (7)$$

Further it is not difficult to see, either by means of (6), or by direct differentiation that:

$$\frac{d\overline{l^2}}{dT} = \frac{\overline{l^2}}{T} + \tfrac{1}{3}(\bar{v})^2 T. \quad (8)$$

3.23. Value of $\overline{vl}$.

Making use again of 3.2(2) we have:

$$\overline{vl} = \frac{1}{T}\int_{t_0}^{t_0+T} dt \int_{t_0}^{t} dt_1\, v(t)\, v(t_1). \quad (1)$$

By introducing variables t_2, η defined by:

$$t_2 = t - t_0;\ \eta = t - t_1,$$

we find (compare 3.123):

$$\overline{vl} = \frac{1}{T}\int_0^T d\eta \int_\eta^T dt_2\, v(t_0 + t_2 - \eta)\, v(t_0 + t_2) =$$

$$= \int_0^T d\eta \frac{T-\eta}{T} S(\eta), \qquad \text{by } 3.13(13) =$$

$$= \int_0^T d\eta \frac{T-\eta}{T} \sigma(\eta) + \int_0^T d\eta \frac{T-\eta}{T} (\bar{v})^2 =$$

$$= \overline{v'^2}\theta + (\bar{v})^2 \frac{T}{2}. \tag{2}$$

Further, as $y = l + y(t_0)$, we find:

$$\overline{vy} = \overline{vl} + \bar{v}\, y(t_0) = \overline{v'^2}\theta + (\bar{v})^2 \frac{T'}{2} + \bar{v}\, y(t_0). \tag{3}$$

The more important results obtained in sections 3.22—3.23 can be summarised by the formulae:

$$\frac{\overline{l^2}}{T} = \frac{\partial}{\partial T}(\overline{l^2}) = \frac{\partial}{\partial T}(\overline{y^2}) = \overline{vl} = \overline{vy} = \overline{v^2}\theta, \tag{4}$$

which are valid provided $\bar{v} = 0$. When $\bar{v}$=constant, these equalities no longer exist.

3.3. Mean values taken over an interval of the order of the duration of correlation

The mean values considered in the preceding sections all have been calculated over a long period T, which is large in comparison with the duration of correlation. This procedure is perfectly applicable in the case where the function $v(t)$ has a stationary character in the statistical sense, so that time mean values can be treated as constants.

In many cases, however, the assumption of a rigorous stationary statistical character cannot be applied. In particular it may not be applicable to the motion of a single particle, even in the case where the general aspect of the field presents a stationary character, as the particle may wander from one part of the field to another. Mean values referring to a short interval of time in such cases will be much more convenient. It is possible to make use of such mean values, when instead of restricting to the motion of one single particle, we consider a great number of particles, starting simultaneously from the same element of volume, and take the average of the mean values referring to the individual partilces. The mean value of any quantity G then is obtained by means of the formula:

$$\overline{G}_* = \frac{1}{N} \sum_i \frac{1}{T} \int_{t_0}^{t_0+T} dt \, G_i(t), \tag{1}$$

where the index i refers to the various individual particles over which the summation is extended. Burgers has made the hypothesis that in this formula the interval T can be reduced to the upper measure of correlation ϑ, provided the number N is sufficiently large.

Mean values calculated in this way will be distinguished by an asterisk, as indicated in (1).

In order to be able to make use of this procedure it is necessary that there can be found a sufficient number of particles (or other entities) to which the quantity G may refer, and which all can start at the same instant from the same place. In the case of particles suspended in a liquid we may take the particles which at the instant t_0 are found in the same element of volume of the field. This idea, however, cannot be applied to the fluid itself, as there are no different individual elements occupying the same spot simultaneously. To overcome this difficulty we recur to an artifice and assume that the field considered is one of a large set of fields, which in the statistical sense behave in identical fashion; we then follow the simultaneous life histories of the elements of volume which at t_0 start from the same spot in all these fields. In this case the various elements are even completely independent of each other. This method also makes it possible to introduce the notion of probability; we may speak for instance of the probability that the velocity of the element situated between y and $y + dy$ at the instant t has a value between v and $v + dv$, or that the path described by the element in the interval from t to $t + \vartheta$ has a length between l and $l + dl$.

Having introduced the „*short interval group mean value*" $\overline{G}_*$ we define:

$$G' = G - \overline{G}_* . \tag{2}$$

We shall not introduce the assumption of a stationary statistical character of the field in a way analogous to the assumption of section 3.11, but content ourselves with the assumption that in an interval of time of the order ϑ the field changes sufficiently little in order that we may use the formula:

$$\overline{G'_*} = 0. \tag{3}$$

3.31. Mean values of the velocity and correlations

With the aid of formulae 3.3(1), 3.3(2) and 3.3(3) we define a mean value of the velocity v, the fluctuation v', and the mean square values $\overline{v^2}$ and $\overline{v'^2}$.

We further introduce the quantities:

$$S_*(\eta) = \frac{1}{N}\sum_i \frac{1}{T}\int_{t_0}^{t_0+T} dt\, v_i(t)\, v_i(t+\eta), \tag{1}$$

$$\sigma_*(\eta) = \frac{1}{N}\sum_i \frac{1}{T}\int_{t_0}^{t_0+T} dt\, v'_i(t)\, v'_i(t+\eta), \tag{2}$$

which are connected by the equation:

$$S_*(\eta) = \sigma_*(\eta) + (\bar{v}_*)^2. \tag{3}$$

We further have:

$$\sigma_*(o) = (\overline{v'^2})_*. \tag{4}$$

With these quantities we define:

$$\varrho_*(\eta) = \frac{\sigma_*(\eta)}{\sigma_*(o)} \tag{5}$$

$$\theta_* = \int_0^\infty d\eta\, \varrho_*(\eta) = \int_0^\vartheta d\eta\, \varrho_*(\eta). \tag{6}$$

3.32. Alternative method of calculating the new mean values

The value of $\bar{G}_*$, given by formula 3.3(1), must be independent of T, as long as T is at least equal to ϑ and on the other hand is not increased in such a way that it becomes large in comparison to ϑ. This also follows from the assumption mentioned at the end of section 3.3, as indeed the mean value $\bar{G}_*$ with sufficient approximation can be considered as a constant, when the value of t_0 is changed by amounts of the order of ϑ.

Burgers has suggested therefore that we might calculate $\overline{G_*}$ also by means of the formula:

$$\bar{G}_* = \frac{1}{N} \sum_i G_i(t_0 + T), \tag{1}$$

provided $T > \vartheta$. If this suggestion is right the mean value of any quantity can be derived from the instantaneous values of this quantity at the instant $t_0 + T$ (if desired, at the instant $t_0 + \vartheta$) for the particles which at the instant t_0 all started from the same element of the field.

We can apply this procedure to the calculation of the mean velocity and obtain:

$$\bar{v}_* = \frac{1}{N} \sum_i v_i(t_0 + T). \tag{2}$$

Comparison of this result with the formulae of Chapter 1 shows that the mean value calculated in this way is identical with the quantity $\overset{\sqcap}{v}$ as defined by 1.7(7) provided we take $T = \tau$. In this way we arrive at a connection between mean values originally defined as time mean values, and the mean values investigated in Chapter 1.

In general it cannot be assumed that equation (1) will remain valid when T is reduced below ϑ, *e.g.* until zero. The result arrived at in section 1.71, where a difference was found between the two quantities

$\bar{v}$ (average velocity of the particles present in a given element of volume at the instant t_0),

and $\overset{\sqcap}{v}$ (average velocity of the same group of particles at the instant $t + \tau$, after they have suffered displacements individually different for the various particles),

brings to light a case where (1) does not hold with $T = 0$ (this difference between $\bar{v}$ and $\overset{\sqcap}{v}$ is also found in the example treated in Chapter 2, section 2.43).

Such differences are to be expected in particular when $\overset{\sqcap}{v}$ and $\bar{v}$ are small, being in fact only residual effects connected with inhomogeneity of turbulence.

It seems probable that in the calculation of mean square values and of correlations similar differences will not occur, or at any rate will be much smaller. (In particular this may be the case when the mean velocity $\bar{v}_*$ is small in comparison to $\sqrt{(\overline{v^2})_*}$). It will be assumed therefore that the equation:

$$S_*(\eta) = \frac{1}{N} \sum_i v_i(t_0 + T)\, v_i(t_0 + T + \eta) \tag{3}$$

will hold for *all* positive and negative values of T (which are not too large) including $T \lesseqgtr 0$.

3.33. Formulae containing the displacement of the particles

We write:

$$l_i(t) = \int_{t_0}^{t} dt_1\, v_i(t_1), \tag{1}$$

where $t = t_0 + T$, and define:

$$l_* = \frac{1}{N} \sum_i l_i \tag{2}$$

$$l_*^2 = \frac{1}{N} \sum_i l_i^2 . \tag{3}$$

These quantities are simple group mean values and are quite distinct from the time mean values for l and l^2 considered in 3.2—3.23. Provisionally we make no restrictions on the values of T.

Differentiation of (2) with respect to T gives:

$$\frac{dl_*}{dT} = \frac{1}{N} \sum_i v_i(t_0 + T) = v_* . \tag{4}$$

It will be seen that when $T > \vartheta$ the quantity v_* introduced here is identical with $\bar{v}_*$ as defined by 3.32(2). In that case also l_*/T gives the same value according to form. 3.3(1). Hence:

$$\frac{dl_*}{dT} = \frac{l_*}{T} = \bar{v}_* \tag{5}$$

provided $T > \vartheta$.

Differentiation of (3) with respect to T gives:

$$\frac{dl_*^2}{dT} = \frac{2}{N} \sum_i v_i(t_0 + T)\, l_i = 2(vl)_* . \tag{6}$$

Here $(vl)_*$ can be written:

$$(vl)_* = \frac{1}{N}\sum_i \int_0^T dt_1\, v_i(t_0 + T)\, v_i(t_0 + t_1) =$$

$$= \frac{1}{N}\sum_i \int_0^T d\eta\, v_i(t_0 + T)\, v_i(t_0 + T - \eta).$$

When we make use of the assumption contained in 3.32(3), we can write:

$$(vl)_* = \int_0^T d\eta\, S_*(\eta). \tag{7}$$

Application of 3.31(3) then leads to:

$$(vl)_* = (\overline{v'^2})_* \int_0^T d\eta\, \varrho_*(\eta) + (\bar{v}_*)^2\, T = (\overline{v'^2})_*\, \theta_* + (\bar{v}_*)^2\, T, \tag{8}$$

the latter expression being valid only for $T > \vartheta$.

Comparison with the formula of Chapter 1 shows that $(vl)_*$ is identical with $\overline{vl}$.

Integration of eq. (6) next gives:

$$l_*^2 = 2(\overline{v'^2})_* \int_0^T dt \int_0^t d\eta\, \varrho_*(\eta) + (\bar{v}_*)^2\, T^2 =$$

$$= 2(\overline{v'^2})_*\, T\, \theta_* + (\bar{v}_*)^2\, T^2 - 2(\overline{v'^2})_* \int_0^T d\eta\, \eta\, \varrho_*(\eta),$$

provided again $T > \vartheta$. Hence:

$$\frac{l_*^2}{T} = 2(\overline{v'^2})_*\, \theta_* + (\bar{v}_*)^2 T - \frac{2}{T}(\overline{v'^2})_* \int_0^T d\eta\, \eta\, \varrho_*(\eta). \tag{9}$$

It follows that the relation

$$\frac{dl_*^2}{dT} = \frac{l_*^2}{T}$$

cannot be exact [compare 3.22(8)]. It is approximated, however, when $(\bar{v}_*)^2 \ll (\overline{v'^2})_*$ in such a degree that T still can be taken sufficiently large to make the last term of (9) negligible.

In making a comparison with the equations of Chapter 1, we must write:

$$l_* = \overline{l}; \quad l_*^2 = \overline{l^2}; \quad (vl)_* = \overline{vl}. \tag{10}$$

In that Chapter we had deduced:

$$\frac{d\,\overline{l^2}}{d\tau} = \frac{\overline{l^2}}{\tau}$$

[compare 1.7(10)], assuming that τ was large in comparison with the duration of correlation ϑ, while terms of the order τ^2 in the formulae for $\overline{l}$ and $\overline{l^2}$ had been neglected. In the example given in Chapter 2, where the value of $\overline{v^2}$ had been calculated for a special dispersion function, it had been found indeed that $\overline{v^2}$ was large compared to $(\overline{v})^2$, being of the order $(\overline{v})^2/\tau$ or even much larger. In such a case the term $(\overline{v'^2})_*\theta_*$ in (9) above will largely surpass the term $(\bar{v}_*)^2 T$.

3.34. Comparison of the new mean values with the „long range" time mean values of section 3.1—3.23

Although some of the results have already been mentioned, it may be useful to summarize the main points as follows :

I. On the assumptions of 3.3 and 3.32 the mean value $(\bar{v})_*$ defined either by means of 3.3(1) or by 3.32(1) will be considered as identical for our purposes with $\bar{v}$ as defined by means of 3.11(1). In the same sense $(\overline{v'^2})_*$ is identical to $\overline{v'^2}$.

II. Equations 3.33(5) and 3.33(10) consequently give:

$$\overline{v} = \frac{d\overline{l}}{d\tau} = \frac{dl_*}{dT} = \frac{l_*}{T} = \bar{v}_* = \bar{v}. \tag{1}$$

III. The quantity $S_*(\eta)$ defined either by 3.31(1) or by 3.32(3) will be considered as identical with $S(\eta)$ as defined by 3.12(1). The same applies to the quantities σ_*, ϱ_* and θ_* derived from S_*, which are identified resp. with σ, ϱ and θ as derived from S.

IV. Equation 3.23(2) gives:

$$\overline{vl} = \overline{v'^2}\,\theta + (\bar{v})^2\frac{T}{2}, \tag{2}$$

while from 3.33(8)

$$(vl)_* = (\overline{v'^2})_*\,\theta_* + (\bar{v}_*)^2\,T. \tag{3}$$

The difference in the second term on the right hand member is due to the fact that $\overline{vl}$ is a time mean value over the whole interval from 0 to T, whereas $(vl)_*$ exclusively refers to the instant T. Under such conditions where $(\bar{v})^2$ is sufficiently small and T not too large in order that this term can be neglected, we shall have:

$$(vl)_* \cong \overline{vl}. \tag{4}$$

V. Under the same conditions we have:

$$\frac{d\overline{l^2}}{d\tau} = \frac{dl_*^2}{dT} \cong 2\,(vl)_* \cong 2\,\overline{vl}. \tag{5}$$

VI. Comparison of 3.33(9) with eq. (2) above shows that:

$$\frac{l_*^2}{T} = 2\,\overline{vl} - \frac{2\overline{(v'^2)}_*}{T}\int_0^T d\eta\;\eta\;\varrho_*(\eta), \tag{6}$$

where the last term can be neglected provided we can take T sufficiently large in comparison with the correlation measure ϑ. Under this condition we arrive at:

$$\frac{\overline{l^2}}{\tau} = \frac{l_*^2}{T} = 2\,\overline{vl}. \tag{7}$$

Equations (1) and (7) show that the quantities $\overline{v}$ and $\overline{l^2}/\tau$, which occur in the diffusion equations of Chapter 1, upon the assumptions mentioned can be calculated from the „long range" time mean values $\overline{v}$ and $\overline{vl}$. This result will be of importance in the developments of Chapters 5 and 6.

VII. Nevertheless we should not lose sight of the possibility that the substitution of (7) for (6) in certain cases may not be legitimate, as it is not certain that we can always choose T large in comparison with ϑ. It is not to be excluded that this circumstance may have consequences which could invalidate certain results of Chapter 5. It seems difficult, however, to obtain a better approximation than (7) which at the same time is sufficiently definite for use in further work.

Chapter 4

EQUATION OF MOTION FOR A PARTICLE SUSPENDED IN A HOMOGENEOUS FIELD

4.1. Equation of Basset, Boussinesq and Oseen for the motion of a particle in a fluid at rest

In Chapter 1 we have seen that the equation characterizing the diffusion of suspended particles contains terms $\overline{l}/\tau$, $\overline{l^2}/2\tau$ referring to the motion of a particle. We shall now investigate how these values can be related to the state of motion of the liquid in which the particles are embedded, by using data concerning the action of the moving fluid upon the particles. In order to do this we must study the motion of a solid particle under the combined action of the force of gravity and of the field of fluid motion. The particles for simplicity will be considered as spherical and rigid, while the field in spatial respect is assumed to be homogeneous, which means that the velocity vector in this field shall be a function of the time t only, and is independent of the coordinates of the point of application. We shall be concerned only with the motion in the direction of y-axis, which is the direction of gravity. The mutual action between the solid particles, rotation of the particles, and wall effects are left out of consideration.

It is useful to begin by defining the quantities which will occur in the calculations:

- $u(t)$ velocity of the fluid surrounding the moving solid particle;
- $\dot{u}, \ddot{u}$ 1st and 2nd derivatives of u with respect to the time t;
- $v(t)$ velocity of the solid particle under the action of gravity and the action of the moving fluid;
- $\dot{v}, \ddot{v}$ 1st and 2nd derivatives of v with respect to the time t;
- ϱ, ϱ' resp. density of the fluid and density of the particle;
- $\nu = \mu/\varrho$ kinematic viscosity of the fluid;
- a radius of the spherical particle;
- g acceleration of gravity.

The problem of the slow motion of a spherical particle under the influence of gravity in a fluid at rest has been studied by A. B. Basset, J. Boussinesq and C. W. Oseen[1]). The equation of motion in the direction of the axis Oy ($+y$ being directed vertically upwards) is:

$$\frac{4\pi a^3}{3}\varrho'\dot{v} = $$
$$= -\frac{2\pi a^3}{3}\varrho\dot{v} - 6\pi\mu a\left\{v + \frac{a}{\sqrt{\pi\nu}}\int_{t_0}^{t} dt_1 \frac{\dot{v}(t_1)}{\sqrt{t-t_1}}\right\} - \frac{4\pi a^3}{3}(\varrho' - \varrho)g. \quad (1)$$

In deriving this equation it has been assumed that both the spherical particle and the fluid have been at rest until the instant $t = t_0$. When t_0 is taken further and further away in the past, the influence of what may have been the state of motion before t_0 will more and more vanish. Hence in the following formulae we shall usually take $-\infty$ as the lower limit of the integral.

In the second member, the term $-(2\pi a^3/3)\,\varrho\dot{v}$ has the meaning of a surplus of inertia added to the particle, due to the pressures connected with the accelerated motion. In the classical problem of non viscous flow we find the same inertia term $-(2\pi a^3/3)\,\varrho\dot{v}$ in the second member. The term $-6\pi\mu a v$ is the resistance as given by Stokes' formula for a sphere moving steadily in a viscous fluid at rest; the term

$$-\frac{6\pi\mu a^2}{\sqrt{\pi\nu}}\int_{t_0}^{t} dt_1 \frac{\dot{v}(t_1)}{\sqrt{t-t_1}}$$

is a resistance due to the accelerated motion of the particle in a viscous fluid. The last term simply is the force of gravity.

4.2. Equation of motion of the particle in a moving fluid

We can transform the equation of Basset and Boussinesq-Oseen for the motion of a particle in a fluid at rest into an equation of motion for a particle in a fluid moving with a variable velocity by considering the following two cases:

(*a*) First we take the case of a particle moving with a velocity $+(v - u)$ in a fluid at rest;

---

[1]) A. B. Basset, *A treatise on hydrodynamics* (Cambridge 1888), Vol, 2, Ch. V. J. Boussinesq, *Théorie analytique de la chaleur* (Paris, 1903), Vol. 2, p. 224. C. W. Oseen, *Hydrodynamik* (Leipzig 1927), p. 132.

(*b*) Secondly we assume that the whole system (particle + fluid) is endowed with a velocity $+\,u(t)$.

In the first case the forces acting upon the particle have the same expression as that given in the second member of 4.1(1), provided everywhere v is changed into $(v - u)$:

$$-\frac{2\pi a^3}{3}\varrho(\dot{v}-\dot{u}) - 6\pi\mu a\left\{(v-u) + \frac{a}{\sqrt{\pi\nu}}\int\limits_{-\infty}^{t} dt_1\, \frac{\dot{v}(t_1)-\dot{u}(t_1)}{\sqrt{t-t_1}}\right\} - \frac{4\pi a^3}{3}(\varrho'-\varrho)\,g. \qquad (1)$$

When next we superpose a rectilinear velocity u upon the whole system (fluid + particle), which is a function of the time, there will be an acceleration u of the fluid, which requires the presence of a pressure gradient:

$$-\frac{\partial p}{\partial y} = \varrho\dot{u}$$

throughout the whole field. This pressure gradient gives a resultant force upon the spherical particle of magnitude:

$$R_1 = (\text{volume of the particle}).(\text{pressure gradient}) =$$

$$= +\frac{4\pi a^3}{3}\varrho\dot{u}. \qquad (2)$$

By combining we arrive at the case of the motion of a sphere with a veloc ty v in a fluid moving with a velocity u. The forces acting upon this sphere are the sum of the forces given by (1) and (2) respectively. Consequently, the equation of motion of a spherical particle, moving under gravity in a fluid moving with a velocity u, is:

$$\frac{4\pi a^3}{3}\varrho'\dot{v} = \frac{4\pi a^3}{3}\varrho\dot{u} - \frac{2\pi a^3}{3}\varrho(\dot{v}-\dot{u}) - 6\pi\mu a\left[(v-u) + \frac{a}{\sqrt{\pi\nu}}\int\limits_{-\infty}^{t} dt_1\, \frac{\dot{v}(t_1)-\dot{u}(t_1)}{\sqrt{t-t_1}}\right] - \frac{4\pi a^3}{3}g(\varrho'-\varrho). \qquad (3)$$

In the place of

$$\frac{4\pi a^3}{3}\varrho\dot{u} - \frac{2\pi a^3}{3}\varrho(\dot{v}-\dot{u})$$

in the right hand member, we can write

$$2\pi a^3\varrho\dot{u} - \frac{2\pi a^3}{3}\varrho\dot{v}.$$

Here $2\pi a^3 \varrho \dot{u}$ = resultant of the pressures which appear in consequence of the acceleration $\dot{u}$ of the fluid; while $-(2\pi a^3/3)\varrho \dot{v}$ = surplus in inertia caused by the pressures resulting from the acceleration $\dot{v}$ of the particle (equivalent to the inertia of a virtual mass $(2\pi a^3/3)\varrho$ attached to the particle).

The other terms in the right hand member of (3) have the following meanings:

$-6\pi\mu a(v-u)$ = Stokes' linear resistance due to the relative velocity $(v-u)$;

$$-6\pi\mu a \frac{a}{\sqrt{\pi\nu}} \int_{-\infty}^{t} dt_1 \frac{\dot{v}(t_1)-\dot{u}(t_1)}{\sqrt{t-t_1}}$$ = resistance due to the relative acceleration $(\dot{v}-\dot{u})$;

$$-\frac{4\pi a^3}{3}(\varrho'-\varrho)\,g$$ = gravitational force.

In order to simplify the notation, we write eq. (3) in the form:

$$\dot{v}(t)-\beta\,\dot{u}(t)+\alpha\,[v(t)-u(t)+c]+\sqrt{\frac{3\alpha\beta}{\pi}}\int_{-\infty}^{t} dt_1 \frac{\dot{v}(t_1)-\dot{u}(t_1)}{\sqrt{t-t_1}}=0 \quad (4)$$

with:

$$\left.\begin{aligned} \alpha &= \frac{3\nu}{a^2}\frac{3\varrho}{2\varrho'+\varrho} && \text{(dimension: } T^{-1}\text{)}\\ \beta &= \frac{3\varrho}{2\varrho'+\varrho} && (\quad ,, \quad : 1)\\ c &= \frac{2}{9}\frac{ga^2}{\nu}\frac{\varrho'-\varrho}{\varrho} && (\quad ,, \quad : LT^{-1}) \end{aligned}\right\} \quad (5)$$

It is easy to verify that c is the limiting velocity of a sphere falling in a fluid at rest, when the velocity of the fall is slow.

In order to have an idea concerning the order of magnitude of these coefficients, we take the following numerical examples:

(*a*) Sand particles suspended in water with

$$a = 10^{-4} - 10^{-2}\text{ cm},\ \varrho = 1,\ \varrho' = 2{,}5,\ \nu = 0{,}013\text{ cm}^2/\text{sec};$$

then: $\alpha = 2.10^6 - 2.10^2\text{ sec}^{-1}$, $\beta = 0{,}5$, $c = 2{,}5.10^{-4} - 2{,}5$ cm/sec.

(*b*) Water particles suspended in air with

$$a = 10^{-4}\text{ cm},\ \varrho = 1{,}2.\ 10^{-3},\ \varrho' = 1,\ \nu = 0{,}145\text{ cm}^2/\text{sec};$$

then: $\alpha = 7{,}8.10^4\text{ sec}^{-1}$, $\beta \backsimeq 1{,}8.10^{-3}$, $c = 0{,}012$ cm/sec.

4.3. Equation of motion in the simplified form

The integral term in equation 4.2(3) is a small quantity; in order to estimate its magnitude we consider a periodic motion such as:

$$v - u = w = w_0 \sin \omega t.$$

In this case:

$$\dot{v} - \dot{u} = \dot{w} = w_0\, \omega \cos \omega t;$$

hence, when we take $t_0 = -\infty$:

$$\frac{a}{\sqrt{\pi\nu}} \int_{-\infty}^{t} dt_1 \, \frac{\dot{v}(t_1) - \dot{u}(t_1)}{\sqrt{t - t_1}} = \frac{a w_0\, \omega}{\sqrt{\pi\nu}} \int_{-\infty}^{t} dt_1 \, \frac{\cos \omega t_1}{\sqrt{t - t_1}} =$$

$$= \frac{a}{\sqrt{\pi\nu}} w_0\, \omega \int_{0}^{\infty} dt_2 \, \frac{\cos \omega (t - t_2)}{\sqrt{t_2}} = a \sqrt{\frac{\omega}{\nu}} \frac{w_0}{\sqrt{2}} \, (\cos \omega t + \sin \omega t).$$

Hence the integral term is of the order $a\sqrt{\omega/\nu}$ with respect to $(v - u)$; it is a small quantity and will be negligible in certain circumstances. By way of approximation in many cases in its place we may write two correction terms, respectively proportional to $(v - u)$ and to $(\dot{v} - \dot{u})$, so that the equation of motion 4.2(3) is brought into the form:

$$\frac{4\pi a^3}{3} \varrho' \dot{v} = \frac{4\pi a^3}{3} \varrho \dot{u} - \frac{2\pi a^3}{3} \varrho \, (1 + \delta') \, (\dot{v} - \dot{u}) -$$

$$- 6\pi\mu a \, (1 + \delta'') \, (v - u) - \frac{4\pi a^3}{3} g(\varrho' - \varrho) \qquad (1)$$

or in contracted form:

$$\dot{v} + \alpha v = f \qquad (2)$$

with:

$$\left.\begin{aligned} f &= \beta \dot{u} + \alpha u - \alpha c \\ \alpha &= \frac{3\nu}{a^2} \frac{3(1 + \delta'')\varrho}{2\varrho' + (1 + \delta')\varrho} \\ \beta &= \frac{(3 + \delta')\varrho}{2\varrho' + (1 + \delta')\varrho} \\ c &= \frac{2}{9} \frac{g a^2}{\nu} \frac{(\varrho' - \varrho)}{(1 + \delta'')\varrho} \end{aligned}\right\} \qquad (3)$$

Equation (1) is the *simplified form* of the equation of motion, while equation 4.2(4) is the *complete form*. In our investigation we shall apply both forms: the former for a preliminary study and the latter for a definite study.

The equation of motion in its simplified form

$$\dot{v} + \alpha v = f$$

can easily be integrated. With the condition $v_0 = u_0$ at the initial instant t_0 its solution is:

$$v = u_0 e^{-\alpha(t-t_0)} + \int_{t_0}^{t} dt_1\, f(t_1)\, e^{-\alpha(t-t_1)}, \tag{4}$$

or in function of u:

$$v = (1-\beta)\, e^{-\alpha(t-t_0)}\, u_0 + \beta u + \alpha(1-\beta) \int_{t_0}^{t} dt_1\, e^{-\alpha(t-t_1)} u(t_1) - \\ - c\,\{1 - e^{-\alpha(t-t_0)}\}. \tag{5}$$

When we put $t_0 = -\infty$ in order to eliminate the influence of what may have been the state of motion before t_0, as stated in 4.1, we can write:

$$v(t) = \int_{-\infty}^{t} dt_1\, f(t_1)\, e^{-\alpha(t-t_1)}, \tag{6}$$

or

$$v(t) = \beta u(t) + \alpha(1-\beta) \int_{-\infty}^{t} dt_1\, e^{-\alpha(t-t_1)}\, u(t_1) - c. \tag{7}$$

4.4. Integration of the equation of motion in its complete form 4.2(4)

The equation of motion to be integrated is the integro-differential equation 4.2(4):

$$\dot{v}(t) - \beta \dot{u}(t) + \alpha[v(t) - u(t) + c] + \sqrt{\frac{3\alpha\beta}{\pi}} \int_{t_0}^{t} dt_1 \frac{\dot{v}(t_1) - \dot{u}(t_1)}{\sqrt{t-t_1}} = 0.$$

Here we have retained t_0 as the lower limit of integration instead of $-\infty$.

To simplify the writing, we put:

$$\left.\begin{aligned} w &= v - u \\ \varphi &= -(1-\beta)\, \dot{u} - \alpha c \\ b &= \sqrt{\frac{3\alpha\beta}{\pi}} \end{aligned}\right\} \tag{1}$$

then at the instant t the motion fulfils the equation:

$$\dot{w}(t) + \alpha w(t) + b \int_{t_0}^{t} dt_1 \frac{\dot{w}(t_1)}{\sqrt{t-t_1}} = \varphi(t). \tag{2}$$

The integro-differential equation with derivatives of the first order will be reduced to a linear differential equation of the second order, the integral being eliminated by this operation. In outline the procedure is as follows: we differentiate (2) once with respect to t; as will be shown below, we then obtain at the instant t:

$$\ddot{w}(t) + \alpha\dot{w}(t) + b\,\frac{\dot{w}(t_0)}{\sqrt{t - t_0}} + b\int_{t_0}^{t} dt_1\,\frac{\ddot{w}(t_1)}{\sqrt{t - t_1}} = \dot{\varphi}(t), \qquad (3)$$

and at the instant t_2:

$$\ddot{w}(t_2) + \alpha\dot{w}(t_2) + b\,\frac{\dot{w}(t_0)}{\sqrt{t_2 - t_1}} + b\int_{t_0}^{t_2} dt_1\,\frac{\ddot{w}(t_1)}{\sqrt{t_2 - t_1}} = \dot{\varphi}(t_2). \qquad (4)$$

Taking the value of the following integral as basis:

$$\int_{t_1}^{t} dt_2\,\frac{1}{\sqrt{(t - t_2)\,(t_2 - t_1)}} = \pi,$$

and applying the following operation to the integral term of equation (4):

$$\int_{t_0}^{t} dt_2\,\frac{1}{\sqrt{t - t_2}}\int_{t_0}^{t_2} dt_1\,\frac{\ddot{w}(t_1)}{\sqrt{t_2 - t_1}} = \pi\,[\dot{w}(t) - \dot{w}(t_0)],$$

we can reduce equation (4), which contains an integral over $\ddot{w}$, into an equation which contains an integral over $\dot{w}$. By comparing this new equation with (2) it becomes possible to eliminate the integral term, so that we obtain an ordinary linear differential equation of the 2nd order. Let us now study this process in detail.

Before effectuating the differentiation we put $t - t_1 = z^2$ in order to prevent the singularity which would occur when $t_1 = t$. The derivative with respect to t of the integral in the left hand member of (2) is:

$$\frac{d}{dt}\int_{t_0}^{t} dt_1\,\frac{\dot{w}(t_1)}{\sqrt{t - t_1}} = 2\,\frac{d}{dt}\int_{0}^{\sqrt{t-t_0}} dz\,\dot{w}(t - z^2) = \left.\frac{\dot{w}(t - z^2)}{\sqrt{t - t_0}}\right|_{z=\sqrt{t-t_0}} +$$

$$+\,2\int_{0}^{\sqrt{t-t_0}} dz\,\ddot{w}(t - z^2) = \frac{\dot{w}(t_0)}{\sqrt{t - t_0}} + \int_{t_0}^{t} dt_1\,\frac{\ddot{w}(t_1)}{\sqrt{t - t_1}}\,.$$

By applying this rule we obtain:

$$\ddot{w}(t) + \alpha\dot{w}(t) + b\frac{\dot{w}(t_0)}{\sqrt{t-t_0}} + b\int_{t_0}^{t} dt_1 \frac{\ddot{w}(t_1)}{\sqrt{t-t_1}} = \dot{\varphi}(t). \tag{3}$$

This formula is fulfilled by the motion at the instant t. The motion at the instant t_2 will fulfil:

$$\ddot{w}(t_2) + \alpha\dot{w}(t_2) + b\frac{\dot{w}(t_0)}{\sqrt{t_2-t_0}} + b\int_{t_0}^{t_2} dt_1 \frac{\ddot{w}(t_1)}{\sqrt{t_2-t_1}} = \dot{\varphi}(t_2). \tag{4}$$

Multiply every term of (4) by $dt_2 . b(t-t_2)^{-\frac{1}{2}}$, integrate with respect to t_2 between the limits t_0 and t and substract from (3); we then obtain:

$$\ddot{w}(t) + \alpha\dot{w}(t) + b\frac{\dot{w}(t_0)}{\sqrt{t-t_0}} - \alpha b\int_{t_0}^{t} dt_2 \frac{\dot{w}(t_2)}{\sqrt{t-t_2}} -$$

$$-b^2\dot{w}(t_0)\int_{t_0}^{t} dt_2 \frac{1}{\sqrt{(t-t_2)(t_2-t_0)}} - b^2\int_{t_0}^{t} dt_2 \int_{t_0}^{t_2} dt_1 \frac{\ddot{w}(t_1)}{\sqrt{(t_2-t_1)(t-t_2)}} =$$

$$= \dot{\varphi}(t) - b\int_{t_0}^{t} dt_2 \frac{\dot{\varphi}(t_2)}{\sqrt{(t-t_2)}}. \tag{5}$$

The second integral of the left hand member of this equation has the value π. In the double integral the order of the integration can be inverted:

$$\int_{t_0}^{t} dt_2 \int_{t_0}^{t_2} dt_1 \frac{\ddot{w}(t_1)}{\sqrt{(t_2-t_1)(t-t_2)}} = \int_{t_0}^{t} dt_1 \int_{t_1}^{t} dt_2 \frac{\ddot{w}(t_1)}{\sqrt{(t_2-t_1)(t-t_2)}} =$$

$$= \int_{t_0}^{t} dt_1\, \ddot{w}(t_1) \int_{t_1}^{t} dt_2 \frac{1}{\sqrt{(t_2-t_1)(t-t_2)}}.$$

The last integral with respect to t_2 has the value π. Hence the value of the double integral becomes:

$$\int_{t_0}^{t} dt_2 \int_{t_0}^{t_2} dt_1 \frac{\ddot{w}(t_1)}{\sqrt{(t_2-t_1)(t-t_2)}} = \pi[\dot{w}(t) - \dot{w}(t_0)].$$

Substituting this into (5) we obtain:

$$\ddot{w}(t) + \alpha\dot{w}(t) + b\frac{\dot{w}(t_0)}{\sqrt{t-t_0}} - \alpha b\int_{t_0}^{t} dt_2 \frac{\dot{w}(t_2)}{\sqrt{t-t_2}} - \pi b^2 \dot{w}(t) =$$

$$= \dot{\varphi}(t) - b\int_{t_0}^{t} dt_2 \frac{\dot{\psi}(t_2)}{\sqrt{t-t_2}}. \tag{6}$$

Let us replace the integral

$$\int_{t_0}^{t} dt_2 \frac{\dot{w}(t_2)}{\sqrt{t-t_2}}$$

in this equation by its value derived from (2). Then (6) becomes:

$$\ddot{w}(t) + 2\alpha\dot{w}(t) + \alpha^2 w(t) - \alpha\varphi(t) - \pi b^2 \dot{w}(t) + b\frac{\dot{w}(t_0)}{\sqrt{t-t_0}} =$$

$$= \dot{\varphi}(t) - b\int_{t_0}^{t} dt_2 \frac{\dot{\varphi}(t_2)}{\sqrt{t-t_2}} \tag{7}$$

or with $t_0 = \infty$:

$$\ddot{w}(t)+2\alpha\left(1-\frac{3\beta}{2}\right)\dot{w}(t) + \alpha^2 w(t) = \dot{\varphi}(t)+\alpha\varphi(t) - b\int_{-\infty}^{t} dt_2 \frac{\dot{\varphi}(t_2)}{\sqrt{t-t_2}}. \tag{8}$$

This is the second order differential equation for w.

We replace w by $(v-u)$, and φ by its value $-(1-\beta)\dot{u} - \alpha c$; for the sake of brevity we introduce a function F and the following coefficients:

$$F(t) = -\alpha_0 c + \alpha_0 u(t) + \alpha_1 \dot{u}(t) + \alpha_2 \ddot{u}(t) - \alpha_3 \int_0^{\infty} dt_3 \frac{\ddot{u}(t-t_3)}{\sqrt{t_3}} \tag{9}$$

$$\left.\begin{aligned} \alpha_0 &= \alpha^2 \\ \alpha_1 &= \alpha(1-2\beta) \\ \alpha_2 &= \beta \\ \alpha_3 &= \sqrt{\frac{3\alpha\beta}{\pi}}(\beta-1) \\ k &= \alpha\left(1-\frac{3\beta}{2}\right) \\ \omega^2 &= \alpha^2 - k^2 = 3\alpha^2\beta\left(1-\frac{3\beta}{4}\right) \end{aligned}\right\} \tag{10}$$

Solving for v, we finally obtain the equation of motion in the form:

$$\ddot{v} + 2k\dot{v} + (k^2+\omega^2)\,v = F. \tag{11}$$

In this equation F is a known function of the time t, determined by the field u. Equation (11) is the equation of motion in the form of a linear differential equation of the 2nd order with a variable second member. We can now obtain the velocity v of the particle explicitly by writing down the solution of this differential equation:

$$v(t) = \frac{1}{\omega} \int_{-\infty}^{t} dt_1 \, e^{-k(t-t_1)} \sin \omega (t - t_1) \, F(t_1) \tag{12}$$

or putting η for $t - t_1$:

$$v(t) = \frac{1}{\omega} \int_{0}^{\infty} d\eta \, e^{-k\eta} \sin \omega\eta \, F(t - \eta). \tag{13}$$

This is the solution of the equation of motion in its complete form, the function F and the coefficients k, ω being given by (10).

4.5. Periodic motion

In order to investigate the motion of a spherical particle in a periodic field, we start from the equation of motion in the complete form 4.2(4); for the sake of simplicity we shall omit the gravitational force. When a spherical partilce is under the action of a given periodic field $u = A\, e^{i\omega t}$, it will vibrate with the same frequency. Let its velocity be represented by $v = B\, e^{i\omega t}$, A and B being two coefficients which can be real or complex. The phase difference between v and u is determined by the argument of B/A, and the ratio of the amplitudes is determined by the magnitude of $| B/A |$.

Before replacing u, v in the left hand member of 4.2(4) by $A\, e^{i\omega t}$ and $B\, e^{i\omega t}$ respectively, let us first pay attention to the integral term. Substituting $\omega(t - t_1) = z$, we find:

$$i\omega \int_{-\infty}^{t} \frac{e^{i\omega t_1} dt_1}{\sqrt{t-t_1}} = i\sqrt{\omega} \int_{0}^{\infty} dz . \frac{e^{i\omega t - iz}}{\sqrt{z}} = i\sqrt{\omega}\, e^{i\omega t} \int_{0}^{\infty} dz . \frac{\cos z - i \sin z}{\sqrt{z}} =$$

$$= i\sqrt{\omega}\, e^{i\omega t} (1 - i) \sqrt{\frac{\pi}{2}} = e^{i\omega t} \sqrt{\pi i \omega}.$$

Hence eq. 4.2(4) becomes:

$$i\omega B - i\omega\beta A + \alpha(B - A) + \sqrt{3i\alpha\beta\omega}\, (B - A) = 0,$$

from which:

$$\frac{B}{A} = \frac{\alpha + i\omega\beta + \sqrt{3i\alpha\beta\omega}}{\alpha + i\omega \ + \sqrt{3i\alpha\beta\omega}}, \tag{1}$$

which also can be written as:

$$\frac{B}{A}=1-\frac{i\omega(1-\beta)}{\alpha+i\omega+\sqrt{3i\alpha\beta\omega}}=1-\frac{2(\varrho'-\varrho)}{(2\varrho'+\varrho)-9\varrho\left(\frac{i\nu}{a^2\omega}+i\sqrt{\frac{i\nu}{a^2\omega}}\right)}. \tag{2}$$

From this expression we see that:

a) in a viscous fluid, $B/A = 1$ for $\omega = 0$ (frequency tending to zero) and $B/A = \beta$ for $\omega = \infty$ (frequency infinitely great), where $\beta \gtrless 1$ according to $\varrho' \lessgtr \varrho$;

b) in a non-viscous fluid ($\nu = 0$), we have: $B/A = \beta$;

c) for liquid particles suspended in air ($\varrho' \gg \varrho$) we have: for very coarse particles ($a \to \infty$), $B/A = \beta = 3\varrho/(2\varrho' + \varrho) \to 0$, and for very fine particles ($a \to 0$), $B/A \to 1$.

The present investigation of the motion of spherical particles in a viscous periodic field is related to the investigation of the influence of sound waves on scattering particles suspended in a viscous medium. In this connection H. Lamb has obtained the following formula for the ratio of the displacement of a spherical particle to that of the air [1]):

$$-\frac{\sigma\xi}{k}=1-\frac{\varrho'-3\varrho\,\psi_1(ka)}{\varrho'-\varrho-\varrho\,ka.f_1'(ka)\,A_1}.$$

We will show that starting from this equation we can again arrive at equation (2). We have from Lamb, p. 655, eq. (8), for great values of the velocity of sound, which is represented by c in Lamb's formula:

$$k^2=\frac{\omega^2}{c^2+{}^4/_3\,i\nu\omega}\cong\frac{\omega^2}{c^2}\,;\quad ka\cong\frac{\omega a}{c}\ll 1.$$

Also from art. 292, p. 504, eq. (7):

$$\psi_n(\zeta)=\frac{1}{1.3\ldots.(2n+1)}\left\{1-\frac{\zeta^2}{2(2n+3)}+\ldots.\right\}$$

so that $\psi_1(ka) = \frac{1}{3}$.

Further from p. 660, eq. (37):

$$-\,ka\,f_1'(ka)\cong\frac{3}{k^3a^3},$$

and from p. 657, eq. (4):

$$A_1\cong-\frac{(3+3ih-h^2a^2)\,k^3a^3}{2h^2a^2}.$$

[1]) H. Lamb, *Hydrodynamics* (Cambridge 1932), Art., 363, form. (36).

Hence:

$$-\frac{\sigma\xi}{k} = 1 - \frac{\varrho' - \varrho}{\varrho' - \varrho - \frac{3}{2}\varrho\,\dfrac{3 + 3ih - h^2a^2}{h^2a^2}} =$$

$$= 1 - \frac{\varrho' - \varrho}{(\varrho' + \varrho/2) - \frac{9}{2}\varrho\left(\dfrac{1}{h^2a^2} + \dfrac{1}{ha}\right)}.$$

As we have:

$$h^2 = \frac{\omega}{i\nu},$$

we find:

$$-\frac{\sigma\xi}{k} = 1 - \frac{2(\varrho' - \varrho)}{(2\varrho' + \varrho) - 9\varrho\left(\dfrac{i\nu}{a^2\omega} + i\sqrt{\dfrac{i\nu}{a^2\omega}}\right)},$$

which is identical with our expression (2) for B/A.

4.6. Motion with non linear resistance

The resistances due to the velocity and to the acceleration, as expressed by the two terms between [ ] in equation 4.2(3), are exact only when the Reynold's number $R_e = 2(v - u)\,a/\nu$ is small. They are not valid for large R_e. As is well known the resistance due to the velocity becomes proportional to $(v - u)^2$ when R_e is large. The resistance due to the acceleration likewise will change for large R_e.

From dimensional analysis we can deduce some conceptions concerning the law of resistance, which must have the form:

$$R = \varrho\, w^2\, a^2 . f\left(\frac{2wa}{\nu}, \frac{\dot{w}a}{w^2}\right). \tag{1}$$

Here $f(2wa/\nu, \dot{w}a/w^2)$ is a function of $2wa/\nu$ and $\dot{w}a/w^2$, both quantities being dimensionless. The second factor indicates the effects of the acceleration. The differential quotient of second order of w and those of higher orders have been neglected. It is often assumed that the above expression for R can be developed into a series of the form:

$$R = \varrho w^2 a^2 f_1\left(\frac{2wa}{\nu}\right) + \varrho w^2 a^2 \frac{\dot{w}a}{w^2} f_2\left(\frac{2wa}{\nu}\right) + \ldots .$$

$$\cong \varrho w^2 a^2 f_1\left(\frac{2wa}{\nu}\right) + \varrho a^3 \dot{w} f_2\left(\frac{2wa}{\nu}\right). \tag{2}$$

If we put:

$$Re = \frac{2wa}{\nu},$$

$$C_1(R_e) = \frac{2}{\pi} f_1\left(\frac{2wa}{\nu}\right),$$

$$C_2(R_e) = \frac{3}{4\pi} f_2\left(\frac{2wa}{\nu}\right),$$

where the functions C_1, C_2 depend on the shape of the suspended particle, the expression for the resistance can be reduced to a more familiar form:

$$R = C_1(R_e) \, . \, \pi a^2 \, . \, \tfrac{1}{2}\varrho w^2 + C_2(R_e) \, . \, \frac{4\pi a^3}{3} \varrho \dot{w}.$$

When w changes sign, the resistance figuring in the first term must also change sign; hence it is more correct to write:

$$R = C_1(R_e) \, . \, \pi a^2 \, . \, \tfrac{1}{2}\varrho \mid w \mid w + C_2(R_e) \, . \, \frac{4\pi a^3}{3} \varrho \dot{w}. \tag{3}$$

In order to obtain the equation of motion of the particle, we add to this expression the terms due to the acceleration of the liquid, and the gravitational force:

$$\frac{4\pi a^3}{3} \varrho' \dot{v} = \frac{4\pi a^3}{3} \varrho \dot{u} - \frac{4\pi a^3}{3} (\varrho' - \varrho)\, g -$$

$$- C_1(R_e) \, . \, \tfrac{1}{2}\pi a^2 \, . \, \varrho \mid v - u \mid (v - u) - C_2(R_e) \, . \, \frac{4\pi a^3}{3} \varrho\, (\dot{v} - \dot{u}). \tag{4}$$

Assembling similar terms, and dividing by $4\pi a^3/3$, the equation becomes:

$$\dot{v}(\varrho' + \varrho C_2) = (\varrho + \varrho C_2)\, \dot{u} - \tfrac{3}{8} C_1 \frac{\varrho}{a} \mid v - u \mid (v - u) - (\varrho' - \varrho)\, g,$$

or:

$$\dot{v} - \frac{\varrho + \varrho C_2}{\varrho' + \varrho C_2} u + \frac{3\varrho C_1}{8(\varrho' + \varrho C_2)} \frac{1}{a} \mid v - u \mid (v - u) + \frac{\varrho' - \varrho}{\varrho' + \varrho C_2} g = 0. \tag{5}$$

Let us introduce the coefficients:

$$\beta' = \frac{\varrho + \varrho C_2}{\varrho' + \varrho C_2}\, ; \; \gamma' = \frac{3\varrho C_1}{8(\varrho' + \varrho C_2)} \frac{1}{a}\, ; \; C = \frac{\varrho' - \varrho}{\varrho' + \varrho C_2} g.$$

It is to be noted that in the case of non-spherical bodies, the coefficient

$$\gamma' = \frac{3\varrho C_1}{8(\varrho' + \varrho C_2)} \frac{1}{a}$$

must be replaced by

$$\frac{\varrho C_1}{2(\varrho' + \varrho C_2)} \cdot \frac{F\ (= \text{area of master section})}{\text{volume}},$$

or some similar expression.

The equation of motion can now be written:

$$\dot{v} - \beta'\dot{u} + \gamma' \mid v - u \mid (v - u) + C = 0. \tag{6}$$

The solution of this differential equation of quadratic form presents excessive mathematical difficulties. It is usually replaced by a linear one:

$$\dot{v} - \beta'\dot{u} + \alpha'(v - u) + C = 0, \tag{7}$$

by choosing conveniently the coefficient α'.

For simplicity we write again $v - u = w$. We will define α' so that the resistances $\alpha' w$ and $\gamma' \mid w \mid w$ give the same negative work during the period T used for the calculation of mean values. This gives:

$$\alpha' = \gamma' \frac{\int_{t_0}^{t_0+T} \mid w \mid w^2 \, dt}{\int_{t_0}^{t_0+T} w^2 \, dt}, \tag{8}$$

or:

$$\alpha' = \gamma' \frac{\overline{(\mid w \mid . w^2)}}{\overline{w^2}}. \tag{9}$$

Hence the ratio α'/γ' depends on the amplitude of w and also on the form of the w-curve (*e.g.* whether the curve is rectangular, more or less harmonic or sharp pointed).

For harmonic periodic motion with:

$$w = A \sin\left(\frac{2\pi t}{T_1}\right),$$

we find:

$$\frac{\alpha'}{\gamma'} = \frac{8}{3\pi} A = \frac{8\sqrt{2}}{3\pi} \sqrt{\overline{w^2}}. \tag{10}$$

By want of detailed data this formula can be used to determine approximately the order of magnitude of α'.

The identification between the equation (4) and the simplified equation of motion studied in 4.3(1) would be complete if we might put:

$$\left.\begin{aligned} C_2 &= \frac{1 + \delta'}{2}, \\ C_1 &= \frac{24}{R_e} (1 + \delta''), \text{ with: } R_e = \frac{2wa}{\nu}. \end{aligned}\right\} \tag{11}$$

Chapter 5

MEAN VALUES CONNECTED WITH THE MOTION OF A PARTICLE, CALCULATED FROM THE EQUATION OF MOTION, BOTH IN THE SIMPLIFIED AND IN THE COMPLETE FORM

5.1. Calculation of mean values based upon the simplified equation of motion

As already done we indicate the velocity of a particle by $v(t)$, and write $u(t)$ for the velocity of the surrounding element of volume of the liquid, which velocity we assume to have been given. When particles should wander from one element of volume into another with a different velocity, the course of the function $u(t)$ will depend both on the nature of the field of motion of the liquid and on the motion of the particles. We evade difficulties by assuming that the elements of the liquid which can be considered as moving with a practically homogeneous velocity, are large not only compared with the dimensions of the suspended particles, but also compared with the paths described by the particles relatively to the liquid.

We shall base the calculations both on the simplified and the complete equation of motion, as deduced in the preceding chapter.

Starting with the first case we take the equation of motion in the form:

$$\dot{v} + \alpha v = f(t), \tag{1}$$

where $f(t)$ is considered as a known function of the time. Later on we shall relate f to the velocity u of the liquid by means of the equation:

$$f = \beta \dot{u} + \alpha(u - c). \tag{2}$$

It is assumed that u and f present a stationary character in the statistical sense. The values of the constants α and β have been given in 4.3(3). The solution of (1) can be written:

$$v = \int_{-\infty}^{t} dt_1\, e^{-\alpha(t-t_1)}\, f(t_1) = \int_{0}^{\infty} d\zeta\, e^{-\alpha\zeta}\, f(t-\zeta). \tag{3}$$

Denote the „long range" time mean value of f by $\bar{f}$, so that

$$\bar{f} = \frac{1}{T}\int_{0}^{T} dt\, f(t)$$

(to shorten notation we replace t_0 by zero, as no derivatives with respect to t_0 will occur in the following pages). We then write:

$$f = f' + \bar{f}. \tag{4}$$

Equation (3) then gives:

$$v = \int_{0}^{\infty} d\zeta\, e^{-\alpha\zeta}\, f'(t-\zeta) + \frac{\bar{f}}{\alpha}. \tag{5}$$

When calculating the time mean value of v, we find that the first term on the right hand member does not give a contribution (as $\overline{f'} = 0$ itself), so that:

$$\bar{v} = \frac{\bar{f}}{\alpha}. \tag{6}$$

Writing:

$$v = v' + \bar{v}, \tag{7}$$

we have:

$$v' = \int_{0}^{\infty} d\zeta\, e^{-\alpha\zeta}\, f'(t-\zeta). \tag{8}$$

We shall mainly be concerned with the consequences of this relation between v' and f', both of which have mean values equal to zero.

The displacement of a particle is given by

$$y = \int_{0}^{t} dt_1\, v(t_1) \tag{9}$$

(to simplify notation we write y instead of l for the path; moreover t_0 has again been taken equal to zero). We can resolve y into two parts, as follows:

$$y = y' + \bar{v}\, t, \tag{10}$$

where

$$y' = \int_0^t dt_1\, v'(t_1). \tag{11}$$

When $\overline{\varphi(o)\varphi(\eta)}$ as before denotes the correlation for a function $\varphi(t)$:

$$\overline{\varphi(o)\varphi(\eta)} = \frac{1}{T}\int_0^T dt\, \varphi(t)\, \varphi(t+\eta),$$

we write:

$$\widehat{\varphi\varphi_t} = \int_0^t d\eta\, \overline{\varphi(o)\, \varphi(\eta)}, \tag{12}$$

and when $\overline{\varphi} = 0$ and the integral is convergent:

$$\widehat{\varphi\varphi} = \int_0^\infty d\eta\, \overline{\varphi(o)\, \varphi(\eta)}. \tag{13}$$

Further we introduce:

$$\widetilde{\varphi\varphi} = \int_0^\infty d\eta\, e^{-\alpha\eta}\, \overline{\varphi(o)\, \varphi(\eta)}. \tag{14}$$

The latter integral is convergent independently of the circumstance whether $\overline{\varphi}$ is equal to zero or not.

5.11. Calculation of $\overline{v'(o)v'(\eta)}$ and related quantities in terms of f'

In the sections 5.11—5.21 we shall temporarily drop the accents and assume that v, y, f and u respectively denote the fluctuating parts of these quantities, so that in particular $\overline{v} = \overline{f} = \overline{u} = 0$. A return to the original notation will be made in 5.3.

In order to calculate $\overline{v(o)\, v(\eta)}$ we write:

$$\overline{v(o)\, v(\eta)} = \frac{1}{T}\int_0^T dt\, v(t)\, v(t+\eta) =$$

$$= \int_0^\infty d\zeta_1 \int_0^\infty d\zeta_2\, e^{-\alpha(\zeta_1+\zeta_2)} \frac{1}{T}\int_0^T dt\, f(t-\zeta_1)\, f(t+\eta-\zeta_2) =$$

$$= \int_0^\infty d\zeta_1 \int_0^\infty d\zeta_2\, e^{-\alpha(\zeta_1+\zeta_2)}\, \overline{f(o)\, f(\eta+\zeta_1-\zeta_2)}. \tag{1}$$

We make a change of variables by means of the formulae:

$$\zeta_2 + \zeta_1 = \sigma; \ \zeta_2 - \zeta_1 = \delta, \tag{2}$$

so that

$$\frac{\partial(\sigma, \delta)}{\partial(\zeta_1, \zeta_2)} = 2.$$

Having regard to fig. 5.11 the double integral then is transformed into:

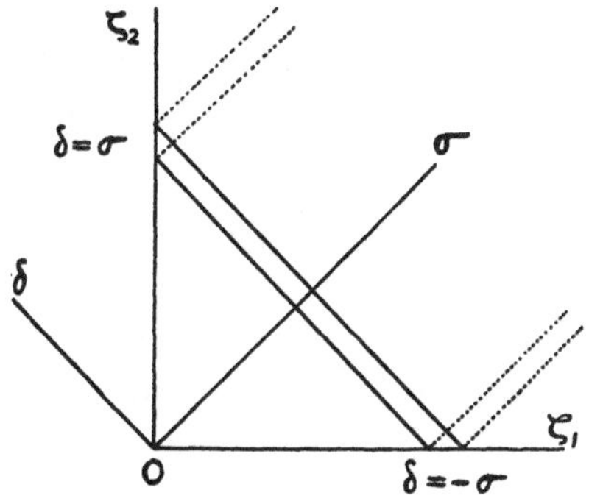

Fig. 5.11.

$$\overline{v(o)\, v(\eta)} = \tfrac{1}{2}\int_0^\infty d\sigma \int_{-\sigma}^{+\sigma} d\delta\, e^{-\alpha\sigma} \{\overline{f(o)\, f(\eta - \delta)}\} =$$

$$= \tfrac{1}{2}\int_0^\infty d\sigma \int_0^\sigma d\delta\, e^{-\alpha\sigma} \{\overline{f(o)\, f(\eta + \delta)} + \overline{f(o)\, f(\eta - \delta)}\} =$$

$$= \tfrac{1}{2}\int_0^\infty d\delta \int_\delta^\infty d\sigma\, e^{-\alpha\sigma} \{\overline{f(o)\, f(\eta + \delta)} + \overline{f(o)\, f(\eta - \delta)}\} =$$

$$= \frac{1}{2\alpha}\int_0^\infty d\delta\, e^{-\alpha\delta} \{\overline{f(o)\, f(\eta + \delta)} + \overline{f(o)\, f(\eta - \delta)}\}. \tag{3}$$

This integral cannot be reduced to a more simple form as long as nothing is given about the function f.

As a special case we take $\eta = 0$, which gives:

$$\overline{v^2} = \frac{1}{\alpha}\int_0^\infty d\delta\, e^{-\alpha\delta}\, \overline{f(o)\, f(\delta)} = \frac{1}{\alpha}\widetilde{ff}\,. \tag{4}$$

The mean square of the velocity consequently depends both upon the correlation existing in the function $f(t)$ and upon the damping coefficient α. As two extreme cases we may take:

a) α so small, that $\alpha\tau \ll 1$, where τ is the upper measure of correlation in f. In this case we can omit the factor $e^{-\alpha\delta}$ in the integral and obtain:

$$\overline{v^2} \cong \frac{1}{\alpha}\int_0^\infty d\delta\, \overline{f(o)\, f(\delta)} = \frac{1}{\alpha}\, \widehat{ff}\ , \tag{5a}$$

b) α so large that the integration in (4) practically extends only over such small values of δ, that we may replace $\overline{f(o)\, f(\delta)}$ by $\overline{f^2}$. In this case:

$$\overline{v^2} \cong \frac{1}{\alpha^2}\, \overline{f^2}. \tag{5b}$$

We next consider:

$$\widehat{vv} = \int_0^\infty d\eta\, \overline{v(o)\, v(\eta)} = \tfrac{1}{2}\int_{-\infty}^{+\infty} d\eta\, \overline{v(o)\, v(\eta)}.$$

Inserting formula (3) the integration with regard to η can be carried out:

$$\widehat{vv} = \frac{1}{4\alpha}\int_0^\infty d\delta\, e^{-\alpha\delta} \int_{-\infty}^{+\infty} d\eta\, \{\overline{f(o)\, f(\eta+\delta)} + \overline{f(o)\, f(\eta-\delta)}\} =$$

$$= \frac{1}{4\alpha}\int_0^\infty d\delta\, e^{-\alpha\delta} . 4\widehat{ff} = \frac{1}{\alpha^2}\, \widehat{ff}. \tag{6}$$

If we introduce the quantities:

$$\theta_v = \frac{\widehat{vv}}{\overline{v^2}}\ ; \quad \theta_f = \frac{\widehat{ff}}{\overline{f^2}} \tag{7}$$

[compare equation 3.13(7)], then from equation (6) and (4) we derive the relation:

$$\theta_v = \frac{1}{\alpha}\frac{\widehat{ff}}{\widetilde{ff}} = \frac{\theta_f}{\alpha}\frac{\overline{f^2}}{\widetilde{ff}}. \tag{8}$$

As we have:

$$\overline{f(o)\, f(\delta)} \leqslant \overline{f^2},$$

it follows that:

$$\widetilde{ff} \leqslant \frac{\overline{f^2}}{\alpha},$$

consequently:

$$\theta_v \geqslant \theta_f. \tag{9}$$

Hence the „duration of correlation" of the function $v(t)$ is usually larger than that of the function $f(t)$, an interesting result. In the extreme cases to which refer equations (5*a*), (5*b*) we find:

$$(\alpha\tau \ll 1) \quad \theta_v \backsimeq \frac{1}{\alpha}, \tag{10a}$$

$$(\alpha\tau \gg 1) \quad \theta_v \backsimeq \theta_f. \tag{10b}$$

5.12. Calculation of $\overline{vf}$

We have immediately:

$$\overline{vf} = \frac{1}{T}\int_0^T dt \int_0^\infty d\zeta\, e^{-\alpha\zeta} f(t)f(t-\zeta) = \int_0^\infty d\zeta\, e^{-\alpha\zeta}\, \overline{f(o)f(\zeta)} = \widetilde{ff}. \tag{1}$$

Comparison with formula 5.11(4) shows that:

$$\overline{vf} = \alpha\overline{v^2}. \tag{2}$$

5.13. Expressions which depend upon the displacement y

With y (which stands for y') given by 5.1(11) we have:

$$\bar{y} = \frac{1}{T}\int_0^T dt \int_0^t dt_1\, v(t_1) \quad \text{or} \quad \int_0^T dt \frac{1}{T}\int_0^t dt_1\, v(t_1).$$

As the limits of the integral with respect to t_1 are 0 and t, instead of 0 and T, we cannot exactly replace it by $\bar{v}$, which would be zero. It will be evident, however, that when T is sufficiently large, the contribution to $\bar{y}$ derived from $(1/T)\int_0^t dt_1\, v(t_1)$ will be, at most, of the order $1/T$. Hence with a certain approximation we may assume:

$$\bar{y} = 0. \tag{1}$$

In the second place:

$$vy = \int_0^t dt_1\, v(t)\, v(t_1) = \int_0^t dt_2\, v(t)\, v(t-t_2).$$

Hence:

$$\overline{vy} = \frac{1}{T}\int_0^T dt \int_0^t dt_2\, v(t)\, v(t-t_2) = \int_0^T dt_2 \frac{1}{T}\int_{t_2}^T dt\, v(t)\, v(t-t_2).$$

In consequence of the corollary given in section 3.122, this can be approximated by:

$$\overline{vy} = \int_0^T dt_2\, \overline{v(o)\, v(t_2)} = \widehat{vv}. \tag{2}$$

Hence by 5.11(6):

$$\overline{vy} = \frac{1}{\alpha^2}\widehat{ff}. \tag{3}$$

It is useful to prove this result in a different way. From 5.1(8) we have:

$$\overline{vy} = \frac{1}{T}\int_0^T dt \int_0^t dt_1 \int_0^\infty d\zeta_1 \int_0^\infty d\zeta_2\, e^{-\alpha(\zeta_1+\zeta_2)}\, f(t-\zeta_1)\, f(t_1-\zeta_2) =$$

$$= \int_0^\infty d\zeta_1 \int_0^\infty d\zeta_2\, e^{-\alpha(\zeta_1+\zeta_2)} \frac{1}{T}\int_0^T dt \int_0^t dt_1\, f(t-\zeta_1)\, f(t_1-\zeta_2). \tag{4}$$

The second double integral can be written:

$$\frac{1}{T}\int_0^T dt \int_0^t dt_2\, f(t-\zeta_1)\, f(t-t_2-\zeta_2) =$$

$$= \frac{1}{T}\int_0^T dt_2 \int_{t_2}^T dt\, f(t-\zeta_1)\, f(t-t_2-\zeta_2).$$

In consequence of the presence of the factor $e^{-\alpha(\zeta_1+\zeta_2)}$ in the complete expression (4) the important domain of values of ζ_1 and ζ_2 is limited and is of the order $1/\alpha$. When T can be chosen such that αT is a large quantity we may again apply corollary 3.122, and the double integral reduces to:

$$\int_0^\infty dt_2\, \overline{f(o)\, f(\zeta_1-\zeta_2-t_2)}.$$

Now by applying the same transformation as was used in the

reduction of 5.11(1) we find:

$$\overline{vy} = \int_0^\infty d\zeta_1 \int_0^\infty d\zeta_2 \, e^{-\alpha(\zeta_1+\zeta_2)} \int_0^\infty dt_2 \, \overline{f(o)\, f(\zeta_1 - \zeta_2 - t_2)} =$$

$$= \tfrac{1}{2} \int_0^\infty d\sigma \int_{-\sigma}^{+\sigma} d\delta \, e^{-\alpha\sigma} \int_0^\infty dt_2 \, \overline{f(o)\, f(\delta - t_2)} =$$

$$= \tfrac{1}{2} \int_0^\infty d\sigma \int_0^\sigma d\delta \, e^{-\alpha\sigma} \int_0^\infty dt_2 \{\overline{f(o)\, f(\delta + t_2)} + \overline{f(o)\, f(\delta - t_2)}\} =$$

$$= \tfrac{1}{2} \int_0^\infty d\sigma \int_0^\sigma d\delta \, e^{-\alpha\sigma} \int_{-\infty}^{+\infty} dt_2 \, \overline{f(o)\, f(\delta + t_2)} =$$

$$= \int_0^\infty d\sigma \int_0^\sigma d\delta \, e^{-\alpha\sigma} \, \widehat{ff} = \frac{1}{\alpha^2} \, \widehat{ff}. \tag{5}$$

5.14. Calculation of $\overline{fy}$

We finally consider the quantity:

$$\overline{fy} = \frac{1}{T} \int_0^T dt \int_0^t dt_1 \int_0^\infty d\zeta \, e^{-\alpha\zeta} f(t_1 - \zeta)\, f(t) =$$

$$= \int_0^\infty d\zeta \, e^{-\alpha\zeta} \frac{1}{T} \int_0^T dt \int_0^t dt_2 \, f(t)\, f(t - t_2 - \zeta).$$

By means of the argument already used we have:

$$\frac{1}{T} \int_0^T dt \int_0^t dt_2 \, f(t)\, f(t - t_2 - \zeta) = \int_0^\infty dt_2 \, \overline{f(o)\, f(t_2 + \zeta)}.$$

Hence:

$$\overline{fy} = \int_0^\infty d\zeta \, e^{-\alpha\zeta} \int_0^\infty dt_2 \, \overline{f(o)\, f(t_2 + \zeta)}. \tag{1}$$

We write $t_2 + \zeta = t_3$; then:

$$\overline{fy} = \int_0^\infty d\zeta \, e^{-\alpha\zeta} \int_\zeta^\infty dt_3 \, \overline{f(o)\, f(t_3)}.$$

Applying partial integration with respect to ζ we find:

$$\overline{fy} = \frac{1}{\alpha}\int_0^\infty dt_3\, \overline{f(o)\, f(t_3)} - \frac{1}{\alpha}\int_0^\infty d\zeta\, e^{-\alpha\zeta}\, \overline{f(o)\, f(\zeta)} = \frac{\widehat{ff}}{\alpha} - \frac{\widetilde{ff}}{\alpha}. \tag{2}$$

Comparison with formulae 5.11(4) and 5.13(3) leads to the result:

$$\overline{fy} = \alpha\, \overline{vy} - \overline{v^2}. \tag{3}$$

This latter relation can be obtained directly from equation 5.1(1) by multiplying this equation by y and observing that

$$y\dot{v} = \frac{d}{dt}(vy) - v^2,$$

the mean value of $d/dt\,(vy)$ being zero.

It may be remarked that for very small values of α ($\alpha\tau \ll 1$) we obtain from (1):

$$\overline{fy} \cong \int_0^\infty d\zeta \int_0^\infty dt_2\, \overline{f(o)\, f(t_2 + \zeta)} =$$

$$= \int_0^\infty d\zeta \int_\zeta^\infty dt_3\, \overline{f(o)\, f(t_3)} = \int_0^\infty d\zeta\, \zeta\, \overline{f(o)\, f(\zeta)} \tag{4a}$$

For very large values of α on the other hand:

$$\overline{fy} \cong \frac{\widehat{ff}}{\alpha}. \tag{4b}$$

Both values will become small when the duration of correlation τ in the function f is small. Considering equation (3) it is of importance to observe, that whereas for small τ all terms in this equation will decrease, nevertheless *for small values of* α we have:

order of $\overline{vy}$: $\frac{\tau}{\alpha^2}\overline{f^2}$ according to 5.13(3);

order of $\overline{v^2}$: $\frac{\tau}{\alpha}\overline{f^2}$ according to 5.11(4);

order of $\overline{fy}$: $\tau^2\,\overline{f^2}$ according to (4*a*) above.

Hence in this case the terms on the right hand member of equation (3) are much more important than the left hand term, and we find:

$$(\alpha\tau \ll 1) \quad \alpha\, \overline{vy} \cong \overline{v^2}. \tag{5}$$

This also follows directly from a comparison between 5.11(4) and 5.13(3), as for $\alpha\tau \rightarrow 0$ we have $\widehat{ff} \cong \widetilde{ff}$.

5.2. Introduction of the expression for f in function of u

The relation between the full values of f and u has been given by formula 5.1(2). As at present we use the letters f and u to denote the variable parts of these quantities exclusively, we must replace this formula temporarily by:

$$f = \beta\dot{u} + \alpha u. \tag{1}$$

We then have:

$$\overline{f(o)\, f(\eta)} = \beta^2\overline{\dot{u}(o)\, \dot{u}(\eta)} + \alpha\beta\left\{\overline{\dot{u}(o)\, u(\eta)} + \overline{u(o)\, \dot{u}(\eta)}\right\} + \alpha^2\overline{u(o)\, u(\eta)}.$$

Now:

$$\overline{\dot{u}(o)\, u(\eta)} + \overline{u(o)\, \dot{u}(\eta)} = 0,$$

while

$$\overline{\dot{u}(o)\, \dot{u}(\eta)} = -\overline{u(o)\, \ddot{u}(\eta)} = -\frac{d^2}{d\eta^2}\overline{u(o)\, u(\eta)}.$$

Hence:

$$\overline{f(o)\, f(\eta)} = \alpha^2\, \overline{u(o)\, u(\eta)} - \beta^2 \frac{d^2}{d\eta^2}\overline{u(o)\, u(\eta)}. \tag{2}$$

For $\eta = 0$ we find:

$$\overline{f^2} = \alpha^2\, \overline{u^2} + \beta^2\, \overline{\dot{u}^2}, \tag{3}$$

either from (2), or directly from (1). Further we can write:

$$\frac{\overline{\dot{u}^2}}{\overline{u^2}} = -\frac{\varrho_0''(\eta)}{\varrho_1(\eta)} = -\varrho_0''(o) = 2\lim_{\eta=0}\frac{1-\varrho_0(\eta)}{\eta^2},$$

where ϱ_0 is the correlation function for u, and ϱ_1 for $\dot{u}$. We write:

$$\lim_{\eta=0}\frac{1-\varrho_0(\eta)}{\eta^2} = \frac{1}{\zeta^2}; \tag{4}$$

then ζ is the intercept on the η-axis of the parabola osculating thee $\varrho_0(\eta)$ curve at its vertex; it is a quantity of the dimension of a time fulfilling the relation:

$$\zeta \leqslant \vartheta,$$

ϑ being the upper measure of correlation in the u-curve. With the aid of (4) equation (3) can be written:

$$\overline{f^2} = \alpha^2\, \overline{u^2}\left(1 + \frac{2\beta^2}{\alpha^2\zeta^2}\right). \tag{5}$$

From (2) we deduce:

$$\widehat{ff} = \int_0^\infty d\eta \, \overline{f(o)\, f(\eta)} = \alpha^2 \, \widehat{uu} - \beta^2 \frac{d}{d\eta} \, \overline{u(o)\, u(\eta)} \Big|_0^\infty = \alpha^2 \, \widehat{uu}, \tag{6}$$

and:

$$\begin{aligned} \widetilde{ff} &= \int_0^\infty d\eta \, e^{-\alpha\eta} \, \overline{f(o)\, f(\eta)} = \\ &= \alpha^2 \, \widetilde{uu} - \beta^2 \int_0^\infty d\eta \, e^{-\alpha\eta} \frac{d^2}{d\eta^2} \, \overline{u(o)\, u(\eta)} = \\ &= \alpha^2 \, \widetilde{uu} - \alpha\beta^2 \int_0^\infty d\eta \, e^{-\alpha\eta} \frac{d}{d\eta} \, \overline{u(o) u(\eta)} = \\ &= \alpha^2 \, \widetilde{uu} + \alpha\beta^2 \, \overline{u^2} - \alpha^2\beta^2 \, \widetilde{uu} = \\ &= \alpha\beta^2 \, \overline{u^2} + \alpha^2 \, (1 - \beta^2) \, \widetilde{uu}. \end{aligned} \tag{7}$$

Introducing:

$$\theta_f = \frac{\widehat{ff}}{\overline{f^2}}, \quad \theta_u = \frac{\widehat{uu}}{\overline{u^2}}, \tag{8}$$

[compare 5.11 (7)], we have:

$$\theta_f = \frac{\alpha^2 \, \widehat{uu}}{\alpha^2 \, \overline{u^2} + \beta^2 \, \overline{\dot{u}^2}} = \frac{\theta_u}{1 + \dfrac{2\beta^2}{\alpha^2\zeta^2}} \, . \tag{9}$$

Hence:

$$\theta_f < \theta_u . \tag{10}$$

5.21. Relation between v and u

Making use of the results obtained in 5.2, equation 5.11 (4) gives:

$$\overline{v^2} = \beta^2 \, \overline{u^2} + \alpha(1 - \beta^2) \, \widetilde{uu}. \tag{1}$$

As $\widetilde{uu} \leqslant \overline{u^2}/\alpha$ (compare the similar formula in f, mentioned in section 5.11), we have:

$$\overline{v^2} \leqslant \overline{u^2}, \tag{2}$$

which result holds good for all values of α and β.
Further from 5.11 (6):

$$\widehat{vv} = \widehat{uu}. \tag{3}$$

Referring to 5.11(9) and 5.2(10) we find:

$$\theta_v > \theta_u > \theta_f. \tag{4}$$

Having regard to 5.13(2) and to the circumstance that an analogous formula will hold good for u, as in the derivation of 5.13(2) no connection with any other function plays a part, we obtain the relation:

$$\overline{vy} = \overline{ux}. \tag{5}$$

In view of the importance of this result it is useful to give a direct demonstration, without passing through the function f. As:

$$\dot{v} + \alpha v = \beta \dot{u} + \alpha u \tag{6}$$

we have:

$$v(t) = \int_0^\infty d\zeta\, e^{-\alpha\zeta} \{\beta \dot{u}(t-\zeta) + \alpha u(t-\zeta)\} =$$

$$= u(t) - (1-\beta) \int_0^\infty d\zeta\, e^{-\alpha\zeta}\, \dot{u}(t-\zeta), \tag{7}$$

and further by integration with respect to t (limits 0 and t)

$$y(t) = x(t) - (1-\beta) \int_0^\infty d\zeta\, e^{-\alpha\zeta} \{u(t-\zeta) - u(-\zeta)\} =$$

$$= x(t) - (1-\beta) \int_0^\infty d\zeta\, e^{-\alpha\zeta}\, u(t-\zeta) + b, \tag{8}$$

where b is a constant. Hence:

$$\overline{vy} = \overline{ux} + b\bar{u} - (1-\beta) \int_0^\infty d\zeta\, e^{-\alpha\zeta} k_1 + (1-\beta)^2 \int_0^\infty d\zeta_1 \int_0^\infty d\zeta_2\, e^{-\alpha(\zeta_1+\zeta_2)} k_2 \tag{9}$$

where k_1 and k_2 are given by:

$$k_1 = \overline{\dot{u}(t-\zeta)\, x(t)} + \overline{u(t)\, u(t-\zeta)} + b\, \overline{\dot{u}(t-\zeta)},$$

$$k_2 = \overline{\dot{u}(t-\zeta_1)\, u(t-\zeta_2)}.$$

Now $\bar{u} = 0$; likewise: $\overline{\dot{u}(t-\zeta)} = 0$; further:

$$\overline{\dot{u}(t-\zeta)\, x(t)} + \overline{u(t)\, u(t-\zeta)} = \frac{d}{dt}\, \overline{u(t-\zeta)\, x(t)} = 0.$$

Hence $$k_1 = 0.$$

Finally:

$$k_2 = \overline{\dot{u}(\zeta_2 - \zeta_1)\, u(o)} = \frac{d}{d\delta}\, \overline{u(\delta)\, u(o)},$$

when $\zeta_2 - \zeta_1 = \delta$. Writing at the same time $\zeta_2 + \zeta_1 = \sigma$, the last term of (8) — leaving apart the factor $(1 - \beta)^2/2$ — reduces to:

$$\int_0^\infty d\sigma \int_{-\sigma}^{+\sigma} d\delta\, e^{-\alpha\sigma} \frac{d}{d\delta}\, \overline{u(\delta)\, u(o)} = \int_0^\infty d\sigma\, e^{-\alpha\sigma} \{\overline{u(\sigma)\, u(o)} - \overline{u(-\sigma)\, u(o)}\} = 0.$$

Consequently there remains:

$$\overline{vy} = \overline{ux}. \tag{10}$$

5.3. Relations between the full values of v and u and derived quantities

We return to our original notation, so that again

$$v = \bar{v} + v', \text{ etc.}$$

From 5.1(2) we deduce:

$$\bar{f} = \alpha\, (\bar{u} - c),$$

so that 5.1(6) gives:

$$\bar{v} = \bar{u} - c, \tag{1}$$

Further with

$$y = y' + \bar{v}t;\ x = x' + \bar{u}t$$

we find:

$$vy = v'y' + \bar{v}y' + v'\bar{v}t + (\bar{v})^2 t,$$

$$ux = u'x' + \bar{u}x' + u'\bar{u}t + (\bar{u})^2 t.$$

When we make use of the relations

$$\overline{y'} = 0,\ \overline{x'} = 0$$

(compare the beginning of section 5.13), we obtain:

$$\overline{vy} = \overline{v'y'} + \tfrac{1}{2}\, (\bar{v})^2 t,$$

$$\overline{ux} = \overline{u'x'} + \tfrac{1}{2}\, (\bar{u})^2 t,$$

and hence on account of 5.21(5), which referred to $\overline{v'y'}$ and $\overline{u'x'}$:

$$\overline{vy} = \overline{ux} + \tfrac{1}{2}\{(\bar{v})^2 - (\bar{u})^2\}\, t = \overline{ux} - \tfrac{1}{2}\{2\bar{u}c - c^2\}\, t. \qquad (2)$$

Under the assumptions made in Chapter 1, where quantities of the order τ have been neglected in calculating $\overline{v}$ and $\overline{vl}$, (reference to these assumptions has again been made in sections 3.33 and 3.34), the second term on the right hand member will be of small importance in comparison with $\overline{ux}$. Hence in dealing with diffusion problems by means of the equations deduced in Chapter 1, we may take:

$$\overline{vy} \backsimeq \overline{ux}. \qquad (3)$$

By means of equations (1) and (3) used together with equation 3.34(1) and 3.34(7) we now are able to connect the equation governing the diffusion of suspended particles with the equation for the diffusion of the elements of volume of the liquid.

5.4. Calculation of the coefficient of diffusion $\overline{vy}$ for the motion regulated by the complete linear equation

As in 5.11—5.21 we consider only the variable parts of v, u and y, and drop the accents. Instead of starting from the simplified equation of motion for the calculation of $\overline{vy}$, as was done in 5.1, we will now start from the complete linear equation of motion studied in 4.4. Its solution has been given in 4.4(13):

$$v(t) = \frac{1}{\omega}\int_0^\infty d\eta\, e^{-k\eta} \sin \omega\eta\, F(t-\eta)$$

the coefficients ω, k, F having been defined in 4.4(9) and 4.4(10).

Hence the mean value $\overline{vy}$ can be written as follows:

$$\overline{vy} = \int_0^\infty d\eta\, \overline{v(o)\, v(\eta)} =$$

$$= \frac{1}{\omega^2}\int_0^\infty d\eta \int_0^\infty d\eta_1 \int_0^\infty d\eta_2\, e^{-k(\eta_1+\eta_2)} \sin \omega\eta_1 \sin \omega\eta_2\, \overline{F(o)\, F(\eta+\eta_1-\eta_2)}.$$

By interchanging η_1, η_2 we arrive at:

$$\overline{vy} = \int_0^\infty d\eta\, \overline{v(o)v(\eta)} =$$

$$= \frac{1}{\omega^2}\int_0^\infty d\eta \int_0^\infty d\eta_2 \int_0^\infty d\eta_1\, e^{-k(\eta_2+\eta_1)} \sin \omega\eta_2 \sin \omega\eta_1\, \overline{F(o)\,F(\eta+\eta_2-\eta_1)}.$$

By putting $\eta_1 - \eta_2 = \eta'$ and adding the two expressions for $\overline{vy}$, we can write:

$$\int_0^\infty d\eta\, \overline{F(o)F(\eta+\eta_1-\eta_2)} + \int_0^\infty d\eta\, \overline{F(o)F(\eta+\eta_2-\eta_1)} =$$

$$= \int_{\eta'}^\infty d\eta\, \overline{F(o)F(\eta)} + \int_{-\eta'}^\infty d\eta\, \overline{F(o)F(\eta)} =$$

$$= \int_{\eta'}^\infty d\eta\, \overline{F(o)F(\eta)} + \int_{-\infty}^{\eta'} d\eta''\, \overline{F(o)F(-\eta'')} =$$

$$= \int_{-\infty}^\infty d\eta\, \overline{F(o)F(\eta)} = 2\int_0^\infty d\eta\, \overline{F(o)F(\eta)} = 2\,\widehat{FF}.$$

Hence:

$$\overline{vy} = \frac{1}{\omega^2}\int_0^\infty d\eta_1 \int_0^\infty d\eta_2\, e^{-k(\eta_1+\eta_2)} \sin \omega\eta_1 \sin \omega\eta_2\, \widehat{FF} =$$

$$= \frac{1}{\omega^2}\int_0^\infty d\eta_1\, e^{-k\eta_1} \sin \omega\eta_1 \int_0^\infty d\eta_2\, e^{-k\eta_2} \sin \omega\eta_2\, \widehat{FF}.$$

As

$$\int_0^\infty d\eta_1\, e^{-k\eta_1} \sin \omega\eta_1 = \frac{\omega}{\alpha^2},$$

with $\alpha^2 = \omega^2 + k^2$, we obtain:

$$\overline{vy} = \frac{1}{\alpha^4}\,\widehat{FF}. \tag{1}$$

5.41. Calculation of $\widehat{FF} = \int_0^\infty d\eta . \overline{F(o) F(\eta)}$

The function F is defined by 4.4(9):

$$F(t) = \alpha^2 u(t) + \alpha_1 \dot{u}(t) + \alpha_2 \ddot{u}(t) - a_3 \int_0^\infty \frac{dt_3}{\sqrt{t_3}} \ddot{u}(t - t_3).$$

Hence the expression $\widehat{FF}$ has the following form [1]:

$$\widehat{FF} = \alpha^4 \int d\eta \, \overline{u(o)\, u(\eta)} + \qquad \mathrm{I}_1$$

$$+ \alpha^2 a_1 \int d\eta \, [\overline{u(o)\, \dot{u}(\eta)} + \overline{\dot{u}(o)\, u(\eta)}] + \qquad \mathrm{I}_2$$

$$+ \alpha_1^2 \int d\eta \, \overline{\dot{u}(o)\, \dot{u}(\eta)} + \qquad \mathrm{I}_3$$

$$+ \alpha^2 \alpha_2 \int d\eta \, [\overline{u(o)\, \ddot{u}(\eta)} + \overline{\ddot{u}(o)\, u(\eta)}] + \qquad \mathrm{I}_4$$

$$+ \alpha_1 \alpha_2 \int d\eta \, [\overline{\dot{u}(o)\, \ddot{u}(\eta)} + \overline{\ddot{u}(o)\, \dot{u}(\eta)}] + \qquad \mathrm{I}_5$$

$$+ \alpha_2^2 \int d\eta \, \overline{\ddot{u}(o)\, \ddot{u}(\eta)} - \qquad \mathrm{I}_6$$

$$- \alpha^2 \alpha_3 \int d\eta \, \frac{dt_3}{\sqrt{t_3}} [\overline{u(\eta)\, \ddot{u}(-t_3)} + \overline{u(o)\, \ddot{u}(\eta - t_3)}] - \qquad \mathrm{I}_7$$

$$- \alpha_1 \alpha_3 \int d\eta \int \frac{dt_3}{\sqrt{t_3}} [\overline{\dot{u}(\eta)\, \ddot{u}(-t_3)} + \overline{\dot{u}(o)\, \ddot{u}(\eta - t_3)}] - \qquad \mathrm{I}_8$$

$$- \alpha_2 \alpha_3 \int d\eta \int \frac{dt_3}{\sqrt{t_3}} [\overline{\ddot{u}(\eta)\, \ddot{u}(-t_3)} + \overline{\ddot{u}(o)\, \ddot{u}(\eta - t_3)}] + \qquad \mathrm{I}_9$$

$$+ \alpha_3^2 \int d\eta \int \frac{dt_3}{\sqrt{t_3}} \int \frac{dt_4}{\sqrt{t_4}} \overline{\ddot{u}(-t_3)\, \ddot{u}(\eta - t_4)}. \qquad \mathrm{I}_{10}$$

These 10 integrals can be calculated by using the properties of correlation, and we shall see that they all vanish except the first one.

a) I_1: Integral I_1 is a known integral:

$$I_1 = \alpha^4 \int d\eta \, \overline{u(o) u(\eta)} = \alpha^4 \, \overline{ux}.$$

b) I_2, I_5: We can write: $\frac{\partial}{\partial t} \overline{[u^{(m)}(t)\, u^{(n)}(t + \eta)]} = 0.$

After differentiation it can be changed into the form:

$$\overline{u^{(m+1)}(t)\, u^{(n)}(t + \eta)} = (-1)\, \overline{u^{(m)}(t)\, u^{(n+1)}(t + \eta)}.$$

[1] The following integrals have their limits of integration (0, ∞).

Again: $$\frac{\partial}{\partial t}\overline{[u^{(m+1)}(t)\, u^{(n)}(t+\eta)]} = 0.$$

from which:

$$\overline{u^{(m+2)}(t)\, u^{(n)}(t+\eta)} = (-1)\,\overline{u^{(m+1)}(t)\, u^{(n+1)}(t+\eta)} = (-1)^2\,\overline{u^{(m)}(t)\, u^{(n+2)}(t+\eta)}$$

Continuation of this process leads to the formula:

$$\overline{u^{(m+p)}(t)\, u^{(n)}(t+\eta)} = (-1)^p\, \overline{u^{(m)}(t)\, u^{(n+p)}(t+\eta)}. \tag{1}$$

Hence I_2 and I_5 vanish.

c) I_3, I_4, I_6: The mean products in these integrals can be changed into the following form:

$$\overline{u^{(m)}(t)\, u^{(m+2n)}(t+\eta)} = (-1)^m \frac{d^{2(m+n)}}{d\eta^{2(m+n)}}\, \overline{u(t)\, u(t+\eta)}.$$

Integration with respect to η gives:

$$\int_0^T d\eta\, \overline{u^{(m)}(t)\, u^{(m+2n)}(t+\eta)} = (-1)^m \frac{d^{2(m+n)-1}}{d\eta^{2(m+n)-1}} R(\eta)\Big|_0^T.$$

Now on account of the symmetry of $R(\eta)$ with respect to $\eta = 0$ it follows that its derivatives of uneven order will vanish at $\eta = 0$, provided of course that $R(\eta)$ remains analytic at $\eta = 0$. This latter property, however, can safely be assumed, as we may obtain a development of $R(\eta)$ into a power series in η by developing $u(t+\eta)$ into a Taylor series. On the other hand we have assumed that the correlation function vanishes for all $\eta > \vartheta$; hence all its derivatives must vanish too. It seems natural to assume that they will vanish without passing through discontinuities. Hence the integrated term on the right hand member of (1) vanishes at both limits, and so

$$I_3 = 0,\ I_4 = 0,\ I_6 = 0.$$

d) I_7, I_8, I_9: The mean products in these integrals can be written in the following form:

$$Q = \int_0^\infty d\eta\, \overline{u^{(m)}(t)\, u^{(m+p)}(t+\eta-t_3)} + (-1)^p \int_0^\infty d\eta\, \overline{u^{(m)}(t)\, u^{(m+p)}(t+\eta+t_3)} =$$

$$= \int_{-t_3}^\infty d\eta\, \overline{u^{(m)}(t)\, u^{(m+p)}(t+\eta)} + (-1)^p \int_{+t_3}^\infty d\eta\, \overline{u^{(m)}(t)\, u^{(m+p)}(t+\eta)} =$$

$$= \int_{-t_3}^\infty d\eta \frac{d^p}{d\eta^p}\, \overline{u^{(m)}(t)\, u^{(m)}(t+\eta)} + (-1)^p \int_{+t_3}^\infty d\eta \frac{d^p}{d\eta^p}\, \overline{u^{(m)}(t)\, u^{(m)}(t+\eta)} =$$

$$= \frac{d^{p-1}}{d\eta^{p-1}}\, \overline{u^{(m)}(t)\, u^{(m)}(t+\eta)}\Big|_{-t_3}^\infty + (-1)^p \frac{d^{p-1}}{d\eta^{p-1}}\, \overline{u^{(m)}(t)\, u^{(m)}(t+\eta)}\Big|_{+t_3}^\infty.$$

In consequence of the fact that $\overline{u^{(m)}(t)\, u^{(m)}(t+\eta)}$ is an even function of η, the derivative has the same absolute value both for $\eta = -t_3$ and for $\eta = +t_3$, while for p uneven the signs are equal, and for p even they are opposite. Hence the terms referring to the limits $-t_3$ and $+t_3$ cancel. As the terms referring to the limits ∞ will vanish according to what has been explained already in *c*), it follows that Q will be zero. Hence integrals I_7, I_8, I_9 vanish.

e) I_{10}: Write:

$$I^* = \int d\eta\, \overline{\ddot{u}(-t_3)\, \ddot{u}(\eta - t_4)} = \int d\eta\, \overline{\ddot{u}(o)\, \ddot{u}(t_3 - t_4 + \eta)} =$$

$$= \overline{\ddot{u}(o)\, \dot{u}(t_3 - t_4 + \eta)}\Big|_0^\infty = -\overline{\ddot{u}(o)\, \dot{u}(t_3 - t_4)}.$$

Now we have:

$$\overline{\ddot{u}(o)\, \dot{u}(\delta)} = \overline{\ddot{u}(-\delta)\, \dot{u}(o)} = -\overline{\dot{u}(-\delta)\, \ddot{u}(o)}.$$

Hence I^* is an odd function in $\delta = t_3 - t_4$, and the integration of I^* symmetrically over the plane (t_3, t_4) will give zero:

$$I_{10} = \alpha_3^2 \int \frac{dt_3}{\sqrt{t_3}} \int \frac{dt_4}{\sqrt{t_4}} I^* = 0.$$

Returning to the expression:

$$\widehat{FF} = \int_0^\infty d\eta . \overline{F(o)F(\eta)} = I_1 + I_2 + \ldots + I_{10},$$

we conclude that among the 10 integrals it contains, the first one only remains, hence:

$$\widehat{FF} = \alpha^4\, \overline{ux}.$$

As from 5.4(1):

$$\overline{vy} = \widehat{FF}/\alpha^4,$$

we finally obtain:

$$\overline{vy} = \overline{ux}. \qquad (1)$$

We see that for a homogeneous field as defined in 4.1 the coefficient of diffusoin $\overline{vy}$ for the particles is equal to the coefficient of diffusion $\overline{ux}$ for the field. This is true when the motion of the particles is regulated by the simplified equation of motion as well as when it is regulated by the complete equation of motion.

5.5. Remarks on Einstein's formula for the Brownian motion

The phenomenon of diffusion of solid particles suspended in a fluid in turbulent motion presents great similarity to the behaviour of colloidal particles in Brownian motion. In the two cases we have to do with a system which is described by statistical laws, and certain methods of reasoning find their application to both of them. In particular this applies to the concept of correlation. Nevertheless certain results seem to come out differently. The usual formula for the Brownian motion is:

$$\overline{y^2} = \tau \frac{R\theta}{N} \frac{1}{3\pi\mu a}, \tag{1}$$

where:

$\overline{y^2}$ = mean square path,
τ = interval of time of observation,
R = gas constant,
θ = absolute temperature,
N = Avogadro's number,
μ = coefficient of absolute viscosity,
a = radius of the spherical particle.

This formula, originally given by Einstein[1]), has been confirmed by other authors, as von Smoluchowski, Langevin, Ornstein, etc. Langevin's deduction is particularly simple. He starts from the equation:

$$\dot{v} + \alpha v = f, \tag{2}$$

where:

v = velocity of the particle;
α = a damping coefficient due to the friction of the liquid opposing the motion of the particle;
f = action of the field surrounding the particle, as a consequence of the impacts of the molecules of the liquid upon the particle. Hence f will be a highly irregular function of the time.

We multiply both members of this equation by y, the coordinate of the particle at the instant t, and take the mean value. This gives:

$$\overline{\dot{v}y} + \alpha\, \overline{vy} = \overline{fy}. \tag{3}$$

---

[1]) Einstein, *Ann. d. Physik.*, **19**, 371, (1906), Von Smoluchowski, *Ann. d. Physik.*, **21**, 756, (1906); Langevin, *Compt. rend.*, **146**, 530, (1908); Ornstein, Versl. Kon. Akad. v. Wet., A'dam, **26**, 1005, (1917).

As:

$$\overline{\dot{v}y} = -\overline{v^2},$$

we may also write:

$$\alpha\,\overline{vy} = \overline{v^2} + \overline{fy}. \tag{4}$$

This equation has the same form as 5.14(3).

Now Langevin takes:

$$\overline{fy} = 0, \tag{5}$$

assuming that there is no correlation between the rapidly and irregularly fluctuating value of f and the coordinate y. Equation (4) then reduces to a relation between $\overline{vy}$ and the mean square velocity $\overline{v^2}$. The latter is calculated from the assumption that the law of the equipartition of energy can be applied to the particles, so that:

$$\tfrac{1}{2}m'\,\overline{v^2} = \frac{R\theta}{2N}, \tag{6}$$

where m' is the mass of the particle. Finally α is deduced from Stokes' formula for the resistance, so that we have:

$$\alpha = \frac{6\pi a\mu}{m'}, \tag{7}$$

(A surplus of inertia due to the acceleration of the liquid surrounding the particle is supposed to be contained in m').

Now the point open to criticism is the assumption (5). Indeed the path of a particle is determined by the forces acting upon it, so that it is not allowed to consider f and y as independent variables. The proper expression for $\overline{fy}$ has been given in 5.14(4a).

Langevin's assumption can be justified, however, for the case of the Brownian motion when we have regard to the estimates given in section 5.14 for the orders of magnitude of the terms occurring in equation 5.14(3). From these estimates it follows that as soon as

$$\alpha\tau \ll 1, \tag{8}$$

we may neglect $\overline{fy}$ in comparison with the other terms of the equation. The relation (8) will be satisfied for particles of dimensions many times exceeding those of molecules, as then the duration of correlation in the motion of these particles (determined by $1/\alpha$) will be much larger than the duration of correlation τ in the forces f which are due to molecular collisions.

On the other hand in our case of particles whose motion is due to turbulence of the fluid, $1/\alpha$ and τ will be of the same order and (8) does not hold good. In this case we must keep to 5.14(2)

$$\overline{fy} = \frac{\widehat{ff}}{\alpha} - \frac{\widetilde{ff}}{\alpha}, \tag{9}$$

from which, by means of 5.2(6) and 5.2(7), we obtain:

$$\overline{fy} = -\beta^2\overline{u^2} + \alpha\,\widehat{uu} - \alpha(1-\beta^2)\,\widetilde{uu}. \tag{10}$$

Equation (4) above then gives

$$\alpha\,\overline{vy} = \overline{v^2} - \beta^2\overline{u^2} + \alpha\,\widehat{uu} - \alpha(1-\beta^2)\,\widetilde{uu},$$

which in consequence of 5.21(1) reduces to

$$\alpha\,\overline{vy} = \alpha\,\widehat{uu},$$

or:

$$\overline{vy} = \widehat{uu} = \overline{ux}.$$

Chapter 6

APPLICATION OF THE RESULTS TO THE PROBLEM OF THE DIFFUSION OF PARTICLES SUSPENDED IN A LIQUID IN TURBULENT MOTION

6.1. Introduction

It is the object of this last chapter to illustrate the theoretical formulae by means of some simple examples. Although a full development of the practical aspect of the problem of the transportation of small particles by a current of liquid or by a gas, with a discussion of the results of experimental investigations, falls outside the scope of this thesis, we will make an application to a case of homogeneous turbulence and to that of particles suspended by the turbulent motion of a stream running horizontally.

The diffusion of the particles suspended in the liquid according to the views adopted in this work, is governed by the equation 1.4(4), deduced in Chapter 1:

$$\frac{\partial n}{\partial t} = -\frac{\partial}{\partial y}\left(n \frac{\overline{l}}{\tau}\right) + \frac{\partial^2}{\partial y^2}\left(n \frac{\overline{l^2}}{2\tau}\right). \tag{1}$$

We suppose the particles to experience the influence of gravity, which in the absence of turbulence would give them a constant velocity of fall $-c$. In the sense indicated in section 1.6 this velocity $-c$ is a „peculiar motion", and in calculating the quantity $\overline{l}/\tau$ we must add $-c$ to the amount deduced from the irregular motion alone. These irregular motions are due to the influence of the turbulence of the liquid. As the development of the method to be applied for the calculation of $\overline{l}/\tau$ and $\overline{l^2}/2\tau$ is dispersed over various parts of the previous Chapters, it seems useful to summarize it shortly, as this will give an opportunity of indicating the most important links of the reasoning and also some points where difficulties have been met and hypotheses had to be introduced to bridge gaps that could not be done away with by exact calculation. These difficulties are connected with the circumstance that mean value theorems had to be applied to phenomena which are essentially of a non-stationary character; hence they are inherent to the subject treated.

The underlying idea is that in order to calculate $\overline{l}/\tau$ and $\overline{l^2}/2\tau$ for the particles we must find a connection between these quantities and similar quantities referring to the motion of the elements of volume of the liquid. The basis for obtaining such a connection has been developed in Chapters 4 and 5, where it has been shown how time mean values for a single particle can be related to time mean values for the surrounding liquid. This result is not yet sufficient for the purpose in view, as the quantities $\overline{l}/\tau$ and $\overline{l^2}/2\tau$ occurring in equation (1) above are group mean values (or „ensemble" mean values) referring to the lengths of the paths described (in the same time interval τ) by a number of particles, simultaneously starting from the same spot.

The problem of connecting such group mean values with time mean values for a single particle has been considered in the second part of Chapter 3 (section 3.3). It is there that the most serious difficulty is met with, which concerns the duration to be assigned to the interval τ (we come back to this point in 6.13).

When we take for granted that group mean values can be replaced by time mean values; when further from Chapter 5, equation 5.3(1) and 5.3(3), we deduce that the time mean values of the velocity $\overline{v}$ and those of the product $\overline{vy}$ or $\overline{vl}$ for the particles in the absence of a peculiar motion due to exterior forces are equal to the corresponding quantities $\overline{u}$ and $\overline{ux}$ or $\overline{u\xi}$ for the liquid; and when finally we make use of equations 1.7(11) in order to pass from $\overline{v}$ and $\overline{vl}$ to $\overline{l}/\tau$ and $\overline{l^2}/2\tau$, we arrive at the result that the quantities $\overline{l}/\tau$ and $\overline{l^2}/2\tau$ for the particle must be equal respectively to the corresponding quantities $\overline{\xi}/\tau$ and $\overline{\xi^2}/2\tau$ for the elements of volume of the liquid. This is the point of view that we shall adopt in constructing the illustrating examples (sections 6.2, 6.3).

Now the dispersion of the elements of volume of an incompressible liquid necessarily must be isomeric (see the definition given in section 1.2); hence, according to 1.21(11):

$$\frac{\overline{\xi}}{\tau} = \frac{\partial}{\partial y}\left(\frac{\overline{\xi^2}}{2\tau}\right). \tag{2}$$

The value of $\overline{\xi^2}/2\tau = \overline{u\xi}$ for the elements of the liquid can be calculated from the turbulent friction. According to the mixture length theory of turbulence [1]), the turbulent stress experienced between adjacent layers of a moving liquid is given by:

$$\text{stress} = \varrho(vl') \frac{dU}{dy} \tag{3}$$

where the quantity (vl') is the analogue of our quantity $\overline{u\xi}$ [2]). As the stress can be calculated from the total resistance experienced by the moving liquid, and as usually sufficient empirical or semi-empirical data will be available concerning the velocity gradient dU/dy, we perceive the possibility of obtaining the numerical data necessary for the application of equation (1). Although here we have given attention, in the first place, to the case without peculiar motion, the extension to the case where such a motion is produced by gravity will not bring any serious difficulty (compare what has been said in section 1.6).

Hence it appears possible in this way to reduce the dispersion problem for the particles to the solution of a partial differential equation with fully known coefficients.

At the same time it follows that the dispersion of the particles will be isomeric, like that of the elements of volume of the liquid. This would settle a problem raised in section 1.72, *viz.* whether inequalities in the field of turbulence could produce an unequal distribution of the particle density, by driving particles from regions of strong turbulence towards regions of weak turbulence. The answer would be negative.

It may seem that this result is an immediate consequence of an assumption introduced in the first paragraph of Chapter 5 (section 5.1), *viz.* that the elements of volume of the liquid which can be considered as moving with a practically homogeneous velocity, are large in comparison with the paths described by the particles relatively to the liquid. It looks as if this assumption already implicitly contains the final result, as it implies that the suspended particles should never leave the elements of volume of the liquid in which they

---

[1]) Compare: S. Goldstein, *Modern Developments in Fluid Dynamics*, (Oxford 1938), vol. I, p. 205–206.

[2]) S. Goldstein, *l.c.*, p. 206, eq. (20).

happen to be embedded, and thus could not have a dispersion different from that of these elements. The question presents itself: is this the only conclusion that can be drawn from the preceding chapters?

On closer inspection a number of points call for attention. These will be considered in the following subsections.

6.11. Influence of an inhomogeneous field of motion of the liquid surrounding a particle

The question can be raised whether the equation for the motion of a single particle in a moving fluid, given in section 4.2 for the case of a homogeneous field, can be extended to cases where the field of motion of the liquid is not homogeneous over the whole region to be considered. In equation 4.2(3) occur two terms in which the time variation $\dot{u}$ of the velocity u of the liquid plays a part: an inertia term (rather two terms, which can be combined into a single term $2\pi\varrho a^3\dot{u}$), and the term expressing the frictional resistance. It is not difficult to deduce that in the case of a field where u is a function both of the time and of the coordinate x, the term $2\pi\varrho a^3\dot{u}$ must be replaced by:

$$2\pi\varrho a^3 \left(\frac{\partial u}{\partial t} + u \frac{\partial u}{\partial x}\right).$$

A more intricate problem is the influence of the variation of the velocity in the viscous resistance; no exact calculation has been made, but it is supposed that here must be introduced the change of u with time as experienced by an observer moving with the particle, so that $\dot{u}$ will have to be expressed as

$$\frac{\partial u}{\partial t} + v \frac{\partial u}{\partial x}.$$

Let us retain the notation $\dot{u}$ for this quantity; then it will be possible to write the inertia term in the form:

$$2\pi\varrho a^3 \dot{u} + 2\pi\varrho a^3 \frac{\partial u}{\partial x} (u - v).$$

In this way we come to the conclusion that to the right hand member of equation 4.2(3) there must be added a term:

$$2\pi\varrho a^3 \frac{\partial u}{\partial x} (u - v).$$

When in this term $\partial u/\partial x$ is considered as a given quantity, which in a first approximation perhaps even may be considered as a constant, the new term can be combined with the term $-6\pi\mu a\ (v-u)$ expressing the ordinary resistance according to Stokes' law. We thus arrive at the result that the presence of the additional term will have a similar influence as a change in the coefficient of viscosity in the expression for the Stokes' resistance; and both equation 4.2(4) and the simplified equation 4.3(1) will retain their respective general forms. Hence in this there is no reason to expect that inhomogenity of the field surrounding a particle will materially alter the conclusions of Chapter 5.

6.12. Mean value of the velocity of the particles

The equality between group mean value and time mean value of the particle velocity has been proclaimed in equation 3.34(1). A similar relation will hold good for the velocity of the elements of the liquid. The equality of $\bar{v}$ and $\bar{u}$ in the absence of a peculiar motion of the particles is laid down in equation 5.3(1). There seems to be no reason to suspect these relations.

We must observe that a systematic velocity $\bar{u}$ different from zero for the elements of the liquid is obtained only in the case of inhomogeneous turbulence, in consequence of the condition that the dispersion process for these elements must be isomeric. From equation 1.21(11) we see that in general the value of $\overline{l}$ will be of a smaller order of magnitude than that of $\overline{l^2}$, unless the turbulence changes considerably over distances of the order of the paths of the elements. (We must exclude, however, such a case, as then the applicability of most of our formulae will be greatly impaired). The relation expressed by 1.21(11) means that particles starting from a given level on the average will move over greater distances in the direction of increasing turbulence, than in the opposite direction. The mean value of the velocities of the element, at the instant they pass the given level, on the other hand is zero. Hence there is no reason to expect a systematic deviation between $\bar{v}$ and $\bar{u}$; consequently neither is there any reason to expect a systematic difference between $\bar{v}$ and $\bar{u}$, or between $\overline{l}$ and $\overline{\xi}$.

6.13. The relation between the mean values $\overline{l^2}$ and $\overline{\xi^2}$.

As these quantities are of a more important order of magnitude than $\overline{l}$ and $\overline{\xi}$ they deserve extra attention.

According to 3.33(6) we have:

$$\frac{d\overline{l^2}}{dT} = 2\,\overline{vl} = 2\int_0^T d\eta\,\{v(t_0+T)\,v(t_0+T-\eta)\}_*. \tag{1}$$

A similar formula can be given for $\overline{\xi^2}$.

Now from 5.21(2) we know that $\overline{v^2} \leqslant \overline{u^2}$; and from 5.2(4) we have: $\theta_v > \theta_u$. We may expect that also $\vartheta_v > \vartheta_u$. When T exceeds both these two quantities we shall have:

$$\int_0^T d\eta\,\{v(t_0+T)\,v(t_0+T-\eta)\}_* = \int_0^T d\eta\,\{u(t_0+T)\,u(t_0+T-\eta)\}_*. \tag{2}$$

in analogy to 5.21(5). For values of T smaller than ϑ_v, however, the first integral will be smaller than the second one. Hence:

$$\frac{d\overline{l^2}}{dT} < \frac{d\overline{\xi^2}}{dT} \quad \text{for} \quad T < \vartheta_v, \tag{3}$$

so that in general:

$$\overline{l^2} \leqslant \overline{\xi^2}. \tag{4}$$

The difference will be the more marked as T is smaller; it is only when T is large compared with ϑ_v that the difference becomes relatively unimportant. It is at this point that the duration of the interval T and that of τ, used in forming the quantities $\overline{l}/\tau$ and $\overline{l^2}/2\tau$, become of significance. Our formulae are valid when these intervals can be chosen large in comparison with the duration of correlation ϑ_v; in that case we may use the sign of equality in (4). On the other hand, τ must still remain small enough in order that quantities like $\overline{l}/\tau$ and $\overline{l^2}/2\tau$ are not influenced by the fact that the particles under consideration should have wandered into regions where the state of turbulence is different from that in the region from where they started. It is not certain that we can always choose T and τ in a way fulfilling both requirements.

When it is necessary to take T and τ rather short, we must expect that in (4) there will be inequality between the two quantities, in such a way that (for short intervals of time) the mean square displacement for the particles appears to be somewhat less than that for the elements of volume of the liquid. This is the consequence of the retardation suffered by the particles, which is effective in short intervals of time (in long intervals it makes that they keep a certain course for a longer time than an element of the liquid will do). The consequence of this circumstance seems to be that the dispersion of the particles will remain somewhat less than that of the elements of the liquid. This might result in a smaller tendency for the particles to be driven away from the regions of intensive turbulence, so that, in consequence of the restoring motion expressed by the mean value $\overline{v}$ or by the value of $\overline{l}/\tau$, they can show a tendency to concentrate slightly towards the regions of intense turbulence. This is contrary to what originally had been expected, and we do not venture to give the result as certain.

6.2. Application of the diffusion equation to the case of homogeneous turbulence

We return to equation 6.1(1) and keep to the assumption that the quantities $\overline{l}/\tau$, $\overline{l^2}/2\tau$ for the particles, in the absence of a peculiar motion produced by exterior forces, are equal to the quantities $\overline{\xi}/\tau$, $\overline{\xi^2}/2\tau$ characterizing the turbulent motion of the elements of volume of the liquid. For the latter we write:

$$\frac{\overline{\xi^2}}{2\tau} = \varphi, \tag{1}$$

and we consider φ as the quantity charatcteristic for the turbulence of the liquid. According to 6.1(2) we then have:

$$\frac{\overline{\xi}}{\tau} = \frac{\partial \varphi}{\partial y}. \tag{2}$$

The corresponding quantities for the particles, when from now onward we purposely introduce a peculiar motion with velocity $-c$, as it will be produced by gravity, are given by:

$$\frac{\overline{l^2}}{2\tau} = \varphi; \quad \frac{\overline{l}}{\tau} = \frac{\partial \varphi}{\partial y} - c. \tag{3}$$

The diffusion equation can then be written:

$$\frac{\partial n}{\partial t} = \frac{\partial}{\partial y}\left(\varphi \frac{\partial n}{\partial y}\right) + c\frac{\partial n}{\partial y}. \tag{4}$$

When we consider a state of equilibrium where there is no variation of the concentration with the time, (4) leads to the following equation:

$$\varphi \frac{dn}{dy} + nc = q = \text{constant}, \tag{5}$$

q being the current of transportation of the particles. When moreover we assume that particles are not introduced into the field, neither are taken away, q must obviously be zero. Hence:

$$\varphi \frac{dn}{dy} + nc = 0. \tag{5a}$$

The solution of this equation is:

$$\frac{n}{n_0} = e^{-c \int_{y_0}^{y} dy/\varphi}, \tag{6}$$

where n_0 is the concentration at the level y_0.

In a reservoir where the coefficient of diffusion φ of the turbulent field is artificially kept constant by stirring, the distribution of the particles will follow the exponential law:

$$\frac{n}{n_0} = e^{-c(y-y_0)/\varphi}. \tag{7}$$

Such homogeneous turbulence in water can be realized by means of periodic agitation with a system of parallel grids, as was applied in the experiments of Hunter Rouse[1]). The spacing between the grids is kept constant over the whole depth of the reservoir, in order to maintain a uniform field of turbulence, and thus a constant value of φ. The distribution of the concentration measured in these experiments appeared to be in good agreement with the exponential law.

It would be worth while to know the relation between the mechanical and geometrical factors characterizing the agitation and the diffusive power of the turbulence produced by it. We can presume that the following quantities will be important in this respect:

- d: diameter of the bars of a grid;
- B: horizontal spacing of the bars in a grid (the grid being supposed to be quadratic);
- A: vertical distance between the grids;
- S: amplitude of the agitation;
- $f = \omega/2\pi$: frequency of the motion;
- c_w: coefficient of resistance of a bar, (which quantity will be dependent upon the form of the bar).

Consider the motion of a portion of unit length of a single bar. Its vertical displacement can be written: $y = S \sin \omega t$, and its velocity

[1]) Hunter Rouse, *Fifth Intern. Congres of Applied Mechanics* (Cambridge, Mass., U.S.A., 1938).

will be: $\dot{y} = S\omega \cos \omega t$. At any instant t this portion experiences a resistance:

$$F = \tfrac{1}{2} c_w \varrho \dot{y}^2 d,$$

and the momentum communicated to the liquid in one half period will be:

$$I_{bar} = \frac{\pi}{4} c_w \varrho S^2 \omega d.$$

We imagine that this momentum originates an eddy of diameter δ in the water, moving with an average velocity $2/\pi\, S\omega$. Such an eddy per unit length will require a momentum:

$$I_{eddy} = \frac{2}{\pi} \varrho S\omega \frac{\pi\delta^2}{4}.$$

By equating $I_{bar} = I_{eddy}$, we obtain the eddy diameter:

$$\delta = \sqrt{\frac{\pi}{2} c_w Sd}\,.$$

As the mixing length will be proportional to the eddy diameter, and the velocity has been assumed to be proportional to $S\omega$, the coefficient of diffusion for the region occupied by the eddy can be written as:

$$\varphi_0 = \lambda\, S\omega \sqrt{c_w Sd},$$

where λ is a numerical constant. In order to find the average value of φ over the whole region of the reservoir, we multiply this expression by the factor $2B\delta S/B^2A$, so that we obtain:

$$\varphi = \lambda\sqrt{2\pi}\,.\,c_w\, \omega \frac{S^3 d}{AB}. \tag{8}$$

Hunter Rouse's experiments, however, do not provide data for testing this relation. Some experiments had been undertaken with this object in the Laboratorium voor Aero- en Hydrodynamica der Technische Hogeschool te Delft by J. J. Bouwman, but owing to war conditions they have not been finished.

6.3. Application of the diffusion equation to particles suspended in a current of water flowing horizontally

We suppose that the state of motion, both of the water and of the particles, is a function of the vertical coordinate y alone. The slight inclination of the streamlines in a actual river with a free surface can be neglected in our equations, although, of course, the slope of the surface determines the driving force which must overcome the frictional resistance of the bed and thus at the same time regulates the intensity of the turbulence.

The frictional stress τ acting between two layers of the stream is a linear function of the depth:

$$\tau = \tau_m \left(1 - \frac{y}{h}\right) \tag{1}$$

where τ_m is the maximum value of the stress, which is experienced at the bottom; h is the total depth of the channel, and y is the distance above the bottom. The value of τ_m is connected with the inclination I of the surface by the relation:

$$\tau_m = \varrho\, g\, h\, I. \tag{2}$$

In the case of a stream heavily loaded with suspended material it will be necessary to multiply the right hand member with the factor $(1 + ns)$, where n is the average number of particles per unit volume, while

$$s = \frac{4\pi a^3}{3} \frac{\varrho' - \varrho}{\varrho},$$

ϱ and ϱ' being respectively the density of the liquid and that of the particles (which are supposed to be spherical and have a radius a).

In virtue of equation 6.1(3) we have:

$$\tau = \varrho \varphi \frac{dU}{dy}, \tag{3}$$

from which:

$$\varphi = \frac{\tau_m}{\varrho} \frac{1 - y/h}{dU/dy}. \tag{4}$$

Equation 6.2(7) then gives:

$$\frac{n}{n_0} = \exp\left\{-\frac{c\varrho}{\tau_m} \int_{y_0}^{y} dy \frac{dU/dy}{1 - y/h}\right\}. \tag{5}$$

Let us take, by way of example, von Kármán's formula for the distribution of the velocity:

$$\frac{U_{max} - U}{\sqrt{\tau_m/\varrho}} = \frac{1}{k}\left[\ln \frac{1}{1 - \sqrt{1 - y/h}} - \sqrt{1 - y/h}\right], \tag{6}$$

where k is a numerical constant approximately equal to 0,36. The coefficient of diffusion φ then comes out as:

$$\varphi = 2kh \sqrt{\frac{\tau_m}{\varrho}} (1 - y/h) (1 - \sqrt{1 - y/h}). \tag{7}$$

which expression presents a maximum at $y/h = 5/9$. Substituting this into (5) the following result is obtained:

$$\frac{n}{n_0} = \left[\frac{1 - \sqrt{1 - y_0/h}}{\sqrt{1 - y_0/h}}\right]^K \cdot \left[\frac{\sqrt{1 - y/h}}{1 - \sqrt{1 - y/h}}\right]^K, \tag{8}$$

where:

$$K = \frac{c}{k} \sqrt{\frac{\varrho}{\tau_m}}. \tag{8a}$$

A somewhat simpler formula for the velocity distribution has been given by Prandtl:

$$\frac{U_{max} - U}{\sqrt{\tau_m/\varrho}} = \frac{1}{k} \ln (2h/y - 1). \tag{9}$$

This gives:

$$\varphi = kh \sqrt{\frac{\tau_m}{\varrho}}\, y/h\, (1 - y/h) (1 - y/2h), \tag{10}$$

which expression presents a maximum at

$$y/h = 1 - \frac{1}{\sqrt{3}}.$$

The formula for n then becomes:

$$\frac{n}{n_0} = \left[\frac{y_0/h\, (1 - y_0/2h)}{(1 - y_0/h)^2}\right]^K \cdot \left[\frac{(1 - y/h)^2}{y/h\, (1 - y/2h)}\right]^K.$$

It is to be remarked that, as formulae (6), (9) for the velocity distribution cannot be applied in the regions near the bottom and near the surface of the water, the formulae for the concentration will not hold good at these regions.

A full theory of the distribution of particles in a channel will request the knowledge of the boundary concentration n_0, especially the concentration at the bottom. For this purpose it will be necessary to study the motion of the particles and of the water in the transition layer where the transportation of materials by traction is ended and the transportation by suspension begins. This forms a special subject of investigation which we must leave aside.

SAMENVATTING

Deze dissertatie **„Problemen betreffende gemiddelden en correlaties bij de beweging van kleine deeltjes welke in een turbulente vloeistof gesuspendeerd zijn"** is ontstaan uit de behoefte om een basis te verkrijgen voor de theoretische behandeling van de diffusie als gevolg van turbulente bewegingen. Wegens de grote omvang van het onderwerp moest worden volstaan met een behandeling van enkele onderdelen. Als zodanig werden van het meeste belang geacht: een analyse van de betrekkingen die ten grondslag liggen aan de diffusievergelijking en de berekening van enige grootheden welke in deze vergelijking optreden en voor de beweging der gesuspendeerde deeltjes karakteristiek zijn, uit grootheden die kenmerkend zijn voor de turbulentie van de vloeistof. Andere problemen, zoals de terugwerking van de deeltjes op het turbulente gedrag van de vloeistof en de randvoorwaarden die in aanmerking zouden moeten worden genomen wanneer de gesuspendeerde materie bv. afkomstig is uit bodemmateriaal van de stroming, moesten daarnaast terzijde worden gelaten. Ter vereenvoudiging is de behandeling overal beperkt tot verplaatsingen in één dimensie, waarvoor gewoonlijk de y-richting is gekozen.

Als grondslag voor de afleiding van de differentiaalvergelijking voor de diffusie der gesuspendeerde deeltjes, is gekozen de theorie van Kolmogoroff. Hierbij wordt uitgegaan van een waarschijnlijkheidsfunctie voor de verplaatsing van een deeltje, in het verdere betoog als „dispersiefunctie" aangeduid. Deze functie:

$$p(t_0, y_0; t, y)\, dy \qquad 1.1(2)\ ^{1)}$$

drukt de waarschijnlijkheid uit dat een deeltje, hetwelk zich ten tijde

[1]) Voor de hoofdstukken en de onderafdelingen daarvan (secties en subsecties) is een decimale nummering toegepast, waarbij het cijfer vóór het punt het betrokken hoofdstuk aangeeft. Vergelijkingen worden geciteerd met het nummer van de betreffende sectie of subsectie, tussen haakjes gevolgd door het nummer der vergelijking in die sectie (of subsectie).

t_0 bevond in het gebied tussen y_0 en $y_0 + dy_0$, ten tijde t zich zal bevinden in het gebied tussen y en $y + dy$. De functie moet daarbij voldoen aan de voorwaarde:

$$\int dy\, p(t_0, y_0; t, y) = 1, \qquad 1.1(3)$$

waarbij de integratie uitgestrekt is te denken over het gehele gebied van waarden, dat voor y is toegelaten. Wanneer $n(t_0, y_0)\, dy_0$ het aantal der deeltjes voorstelt, die zich op het ogenblik t_0 bevinden in het gebied tussen y_0 en $y_0 + dy_0$, dan wordt het aantal deeltjes $n(t, y)\, dy$ welke zich ten tijde t zullen bevinden in het gebied tussen y en $y + dy$ verkregen uit:

$$n(t, y) = \int dy_0\; n(t_0, y_0)\; p(t_0, y_0; t, y). \qquad 1.1(4)$$

De dispersiefunctie moet voldoen aan de volgende integraalvoorwaarde:

$$p(t_0, y_0; t, y) = \int dy''\, p(t_0, y_0; t'', y'')\; p(t'', y''; t, y) \qquad 1.11(3)$$

welke verzekert dat men de verdeling van de deeltjes op het tijdstip t even goed rechtstreeks kan berekenen door middel van de dispersiefunctie $p(t_0, y_0; t, y)$, als in stappen door eerst de diffusie te beschouwen in het tijdsinterval (t, t'') met behulp van de dispersiefunctie $p(t_0, y_0; t'', y'')$, om vervolgens te bezien de diffusie van uit de situatie y'' gedurende het tijdsinterval (t'', t), met behulp van de dispersiefunctie $p(t'', y''; t, y)$; hierbij is t'' een willekeurig tijdstip tussen t_0 en t gelegen.

Voorts worden gedefinieerd de grootheden:

$$\overline{l^m} = \int dy\, (y - y_0)^m\; p(t_0, y_0; t, y). \qquad 1.12(1)$$

K o l m o g o r o f f beperkt zich nu in zijn theorie tot dispersiefuncties, waarvoor bij onbepaalde afname van het tijdsinterval $t - t_0$: (*a*) de grenswaarde $\overline{l}/(t - t_0)$ hetzij tot nul, hetzij tot een eindig bedrag nadert [1]; (*b*) de grenswaarde $\overline{l^2}(t - t_0)$ tot een eindig bedrag nadert; en (*c*) alle grenswaarden $\overline{l^m}/(t - t_0)$ met $m > 2$ tot nul naderen. Dergelijke dispersiefuncties bezitten een aantal eigenschappen, die hen geschikt doen zijn voor de behandeling van fysische vraagstukken, en die het mogelijk maken een partiële differentiaalvergelijking van de 2de orde voor de verdelingsfunctie n af te leiden,

[1] l is de verplaatsing $y - y_0$.

welke dan vervolgens kan dienen voor de behandeling van het diffusieprobleem.

De onderstelling van Kolmogoroff omtrent het bestaan van zekere limieten wanneer $t - t_0$ tot nul nadert, moet evenwel van fysisch standpunt uit als een abstractie worden gezien, welke haar nut heeft omdat zij het afleiden van tal van betrekkingen mogelijk maakt, doch die zich in de werkelijkheid nooit geheel laat realiseren. Bij alle problemen van onregelmatige bewegingen van deeltjes treedt een zg. „correlatieduur" op, welke kan worden beschouwd als een maat voor de tijd, gedurende welke een deeltje een bepaalde beweging als het ware nog enigermate vasthoudt. De beweging gaat eerst haar echt onregelmatig karakter vertonen, wanneer men tijdsintervallen invoert, welke groter zijn, liefst flink wat groter, dan deze correlatieduur. Doet men dit niet, dan zal men tot de conclusie moeten komen dat $\lim \overline{l^2}/2\tau = 0$ wanneer τ streng naar nul gaat. Vandaar dat men bij de fysische toepassing der theorie zich dient af te vragen of de door Kolmogoroff ingevoerde limietwaarden reeds met een redelijke graad van nauwkeurigheid worden bereikt, wanneer het tijdsinterval $t - t_0$ nog niet beneden de correlatieduur is gedaald. Op dit probleem wordt in Hoofdstuk 3 nader ingegaan. Het is bij turbulente bewegingen van een vloeistof niet steeds mogelijk er voor zorg te dragen dat aan de eis omtrent de intervalgrootte wordt voldaan. Dit punt vormt een moeilijkheid in de theorie, die in verschillende gedeelten van de dissertatie telkens weer naar voren treedt.

De behandeling van de eigenschappen der dispersiefunctie en de daaruit afgeleide differentiaalvergelijking voor n vormen het onderwerp van Hoofdstuk 1 [1]). Daarbij is gebleken dat een bepaald type van dispersiefuncties bizondere aandacht verdient. Dit betreft functies welke een verdeling met constante waarde van n onveranderd laten. Hiertoe moet voldaan zijn aan de volgende integraalvoorwaarde:

$$\int dy_0\, p(t_0, y_0; t, y) = 1. \qquad 1.2(1)$$

De bedoelde dispersiefuncties zijn in Hoofdstuk 1 „*isomere dispersiefuncties*" genoemd, daar zij slechts een plaatsverwisseling van deeltjes teweegbrengen, zonder dat de aantallen per volume-eenheid

[1]) Zie ook de publicatie, genoemd in noot 1), blz. 1 van Chapter 1.

veranderen. Bij isomere dispersie blijkt de eigenschap te bestaan:

$$\overline{l} = \frac{d}{dy}\frac{\overline{l^2}}{2} \qquad 1.21(16)$$

welke in vele formules een belangrijke rol speelt.

In Hoofdstuk 1 is voorts aandacht geschonken aan het samengaan van diffusie tengevolge van onregelmatige bewegingen met een bizondere, gerichte beweging tengevolge van de werking van uitwendige krachten, zoals bv. de zwaartekracht, welke de deeltjes wil doen zinken. Waar uit de hierboven vermelde betrekking 1.21(16) volgt dat bij isomere dispersie een systematische verplaatsing van deeltjes, gekenmerkt door een van nul verschillende gemiddelde waarde $\overline{l}$, reeds optreedt zodra de turbulentie niet homogeen is ($\overline{l^2}$ een functie van y), blijkt het onmogelijk de aanwezigheid van een bizondere beweging ten gevolge van een uitwendige kracht te constateren uitsluitend uit waarnemingen over de beweging der deeltjes. Immers, beschikt men alleen over fenomenologische gegevens, dan kan men een dergelijke beweging niet scheiden van een systematische beweging als gevolg van inhomogene turbulentie.

Wegens het eigenaardige karakter van de isomere dispersie en de betekenis welke haar analyse heeft voor het onderwerp van de dissertatie, zijn in Hoofdstuk 2 enkele voorbeelden van dispersiefuncties behandeld, waaraan de in Hoofdstuk 1 ter sprake gekomen eigenschappen konden worden gedemonstreerd. Met enige uitvoerigheid is daarbij in secties 2.3—2.44 een voorbeeld ontwikkeld van een isomere dispersiefunctie van een algemener type, welke behalve de waarschijnlijkheid voor een bepaalde verplaatsing, tevens de waarschijnlijkheid geeft voor het bezitten van een bepaalde snelheid aan het einde der beschouwde periode. Dit maakt het mogelijk belangrijke eigenschappen der dispersiefunctie op vollediger wijze te illustreren, dan het geval is met functies waarin de snelheid geen rol speelt.

De diffusievergelijking, welke in Hoofdstuk 1 is verkregen, heeft de vorm

$$\frac{\partial n}{\partial t} = -\frac{\partial}{\partial y}\left(n\frac{\overline{l}}{\tau}\right) + \frac{\partial^2}{\partial y^2}\left(n\frac{\overline{l^2}}{2\tau}\right). \qquad 1.4(4)$$

Om haar op een bepaald geval te kunnen toepassen, dienen bekend te zijn de grootheden $\overline{l}/\tau$ en $\overline{l^2}/2\tau$ voor de deeltjes welke men wil beschouwen. In het geval van deeltjes gesuspendeerd in een turbulente vloeistof, moeten deze grootheden worden afgeleid uit hetgeen bekend is omtrent de beweging van een enkel deeltje. Daartoe is in Hoofdstuk 3 een algemene analyse gegeven van de eigenschappen van tijdsgemiddelden welke bij de beweging van een deeltje optreden, terwijl in Hoofdstuk 4 de bewegingsvergelijking voor een enkel deeltje, in verband met de beweging van de omgevende vloeistof, is opgesteld en geïntegreerd. Het blijkt in Hoofdstuk 3 dat men bij het analyseren van tijdsgemiddelden voor de beweging van een enkel deeltje onmiddellijk stoot op de zg. „correlatieduur" (welk begrip reeds ter sprake is gekomen in verband met Kolmogoroff's theorie in Hoofdstuk 1); thans kunnen formules worden opgesteld waardoor deze correlatieduur wordt gedefinieerd (zie sectie 3.121). Men vindt dan dat tijdsgemiddelden voor de beweging van een enkel deeltje, willen zij een veilige basis vormen voor verdere beschouwingen, moeten worden genomen over een periode, groot ten opzichte van de correlatieduur. Daar echter de aard van de turbulentie het veelal niet mogelijk maakt voldoende lange perioden toe te passen — immers, wanneer men te doen heeft met een veld met inhomogene turbulentie, kan in een dergelijke periode een deeltje gemakkelijk van uit een bepaald gebied verplaatst zijn naar een gebied, waar de turbulentie een andere intensiteit heeft — moet gezocht worden of men met kortere tijdsduren kan volstaan, wanneer men zich niet tot één deeltje beperkt, doch een groep beschouwt van zich gelijktijdig, zoveel mogelijk onder dezelfde omstandigheden bewegende, deeltjes. Dit probleem is in beschouwing genomen in sectie 3.3—3.34; hierbij wordt een brug geslagen van de zuivere tijdsgemiddelden naar de „groepsgemiddelden", welke optraden in de analyse van Hoofdstuk 1.

Van bizondere betekenis is hierbij nu de betrekking

$$\frac{dl_*^2}{dT} = 2\,(vl)_*, \qquad 3.33(6)$$

waarbij het sterretje aanduidt dat groepsgemiddelden zijn bedoeld. Deze betrekking was reeds in Hoofdstuk 1 gevonden in de vorm:

$$\frac{d\overline{l^2}}{d\tau} = \frac{\overline{l^2}}{\tau} = 2\,\overline{vl}, \qquad 1.7(10) \text{ en } 1.7(11)$$

terwijl zij bij het in Hoofdstuk 2 uitvoerig behandelde voorbeeld een illustratie kon vinden in 2.43(2). De beschouwingen van 3.3—3.34 maken het mogelijk duidelijker aan te geven waardoor de geldigheid van deze formule wordt begrensd.

De in Hoofdstuk 4 opgestelde bewegingsvergelijking voor een enkel deeltje is gevonden door middel van een uitbreiding van een vergelijking, die werd afgeleid door Basset, Boussinesq, en Oseen voor de niet-stationaire beweging van een bol in een vloeistof bij kleine waarden van het getal van Reynolds. Deze vergelijking blijkt een integrodifferentiaalvergelijking van de eerste orde te zijn:

$$\frac{4\pi a^3}{3}\left(\varrho' + \frac{\varrho}{2}\right)\dot{v} =$$

$$= 2\pi a^3 \varrho \dot{u} - 6\pi\mu a\left[(v-u) + \frac{a}{\sqrt{\pi\nu}}\int_{-\infty}^{t} dt_1 \frac{\dot{v}(t_1)-\dot{u}(t_1)}{\sqrt{t-t_1}}\right] - \frac{4\pi a^3}{3} g(\varrho'-\varrho), \qquad 4.2(3)$$

waarin:

a: straal van het bolvormige deeltje;
ϱ': dichtheid van het deeltje;
ϱ: dichtheid van de vloeistof;
μ: viscositeit van de vloeistof;
ν: μ/ϱ;
v: snelheid van het deeltje;
u: snelheid van de vloeistof;
g: versnelling van de zwaartekracht.

Het is gelukt deze vergelijking te transformeren in de volgende gewone differentiaalvergelijking van de 2e orde:

$$\ddot{v} + 2k\dot{v} + (k^2 + \omega^2)\, v = F, \qquad 4.4(11)$$

waarin k en ω bepaalde constante coëfficienten zijn, terwijl F een functie van t is, welke kan worden berekend uit u en de afgeleiden van u naar t.

Uit de volledige vergelijking is ook een vereenvoudigde vergelijking van de eerste orde geconstrueerd:

$$\dot{v} + \alpha v = f, \qquad 4.3(2)$$

waarin de grootheid f optreedt als een op het deeltje werkende kracht, bepaald door:

$$f = \beta\dot{u} + \alpha u - \alpha c, \qquad 4.3(3)$$

met c = valsnelheid; terwijl α en β bepaalde coëfficienten zijn. Aangegeven is door welke benaderingen deze vergelijking uit de volledige vergelijking wordt verkregen. Voor de discussie van een aantal eigenschappen der beweging kan de vereenvoudigde vergelijking een eerste aanloop geven, welke in vele gevallen reeds tot een genoegzaam inzicht voert.

Zowel voor de volledige als voor de vereenvoudigde vergelijking zijn de oplossingen gegeven. Bij de volledige vergelijking is ook het probleem van de periodieke beweging behandeld.

In Hoofdstuk 5 is met behulp van deze oplossingen berekend welk verband bestaat tussen de in Hoofdstuk 3 gedefinieerde tijdsgemiddelden voor de beweging van een deeltje, en overeenkomstige tijdsgemiddelden voor de beweging van de vloeistof welke het deeltje omringt. Deze berekeningen zijn in hoofdzaak uitgevoerd met behulp van de vereenvoudigde vergelijking. Duidt men de verplaatsing van het deeltje aan met y, die van de vloeistof met x, en schrijft men voor de correlatieduur, kenmerkend voor de beweging van het deeltje θ_v, voor de correlatieduur kenmerkend voor de beweging van de vloeistof θ_u, dan is onder meer gevonden:

$$\overline{v^2} \leqslant \overline{u^2}, \qquad 5.21(2)$$

$$\theta_v \geqslant \theta_u, \qquad 5.21(4)$$

$$\overline{vy} = \overline{ux}, \qquad 5.21(5)$$

De laatste dezer betrekkingen is voor het onderwerp van de dissertatie van uitermate groot belang, in verband met de in Hoofdstukken 1 en 3 afgeleide formule:

$$\frac{d\overline{l^2}}{d\tau} = \frac{\overline{l^2}}{\tau} = 2\overline{vl}\ (= 2\overline{vy}).$$

Met het oog daarop is de genoemde betrekking ook bewezen met behulp van de oplossing der volledige bewegingsvergelijking.

Bij de afleiding der betrekkingen 5.21(2), 5.21(4) en 5.21(5) uit de vereenvoudigde vergelijking 4.3(2) is als intermediair opgetreden de op het deeltje werkende kracht f. Incidenteel werd daarbij de volgende formule verkregen:

$$\overline{fy} = \frac{\widehat{ff}}{\alpha} - \frac{\widetilde{ff}}{\alpha}, \qquad 5.14(2)$$

waar

$$\widehat{ff} = \int_0^\infty d\eta\, \overline{f(o)\, f(\eta)}, \qquad \text{vgl. } 5.1(13)$$

$$\widetilde{ff} = \int_0^\infty d\eta\, e^{-\alpha\eta}\, \overline{f(o)\, f(\eta)}, \qquad \text{vgl. } 5.1(14)$$

zie sectie 5.1. De betrekking 5.14(2) is van betekenis, omdat een met $\overline{fy}$ analoge grootheid in Langevin's theorie van de Brownse beweging gelijk aan nul wordt gesteld. In sectie 5.5 is aangetoond dat dit bij de Brownse beweging geoorloofd is op grond van de omstandigheid dat de correlatieduur voor de moleculaire krachten waarvan het deeltje de werking ondervindt (de botsingen der moleculen tegen het deeltje) uitermate klein is tegenover de grootheid $1/\alpha$, welke maatgevend is voor de correlatieduur van de beweging van het deeltje. Bij de diffusie van deeltjes door turbulentie zal men echter in het algemeen een dergelijk verschil tussen de correlatieduren niet mogen verwachten, zodat dan Langevin's onderstelling niet kan worden toegepast.

Tenslotte zijn in Hoofdstuk 6 als voorbeeld een paar zeer elementaire toepassingen gegeven ter illustratie van de wijze waarop men de voor de turbulentie van de vloeistof kenmerkende grootheid $\overline{ux}$ kan berekenen uit gegevens omtrent de schuifspanning en de verdeling van de gemiddelde snelheid; en voorts van het principe volgens hetwelk uit deze grootheid de waarden van $\overline{l}/\tau$ en $\overline{l^2}/2\tau$ voor de beweging van de deeltjes worden afgeleid. Hierbij deed zich tevens de gelegenheid voor om nog kort terug te komen op enkele principes, die aan de theorie ten grondslag zijn gelegd en op de moeilijkheden welke zich daarbij hadden voorgedaan.

Gaarne moge ik hier vermelden dat het verblijf in Nederland en de studie aan de Technische Hoogeschool te Delft mij is mogelijk gemaakt door een studiebeurs van de Academia Sinica, verleend in 1939, waarvoor ik hier mijn oprechte dank uitspreek.

STELLINGEN

I

De onderstelling van K o l m o g o r o f f dat de verhoudingen $\overline{l}/\tau$ en $\overline{l^2}/2\tau$ waarin

$$\overline{l} = \int dl\, l\, P(t, y; \tau, l)$$

$$\overline{l^2} = \int dl\, l^2\, P(t, y; \tau, l)$$

resp. constante waarden zullen benaderen wanneer τ naar nul daalt, is voor fysische verschijnselen niet algemeen juist.

A. K o l m o g o r o f f, *Math. Ann.*, **104**, 45–458, (1931).

II

De onderstelling gemaakt door B u r g e r s, in een artikel over het onderscheid tussen onregelmatige en systematische beweging bij diffusie problemen, dat bij afwezigheid van uitwendige krachten geen systematische beweging zal optreden, is niet juist.

J. M. B u r g e r s, Proc. Acad., Amsterdam, **44**, 344, (1941).

III

De beschouwingen uitgesproken door B u r g e r s in een discussieopmerking bij een voordracht van P r a n d t l, over de verschillende diffusie-coëfficienten zijn niet geheel juist.

J. M. B u r g e r s, in: *Vorträge aus dem Gebiete der Aerodynamik u. verwandter Gebiete,* (Aachen 1929), p. 3.

IV

M i l a t z leidt de formule voor het verband tussen de methodes van O r n s t e i n en S c h o t t k y in de theorie der Brownse beweging af door een bijzondere vorm van de spectrale functie te onderstellen. Deze onderstelling is willekeurig. Het is echter mogelijk deze formule algemeen af te leiden zonder een onderstelling in te voeren, en wel op 2 verschillende manieren: *a*) door de berekening van een integraal van D i r i c h l e t, of *b*) door middel van een F o u r i e r-transformatie van de uitdrukking voor de correlatiefunctie. De afgeleide formule is van belang in de theorie van de mechanische interpretatie van onregelmatige of pseudo-onregelmatige bewegingen.

J. M. W. M i l a t z, *Ned. T. Natuurk.*, **8**, 27, (1941).

V

In een vloeistof met inhomogeen verdeelde turbulentie, moge de gemiddelde snelheid gemeten op een vast punt (b.v. door middel van een hittedraad-anemometer) worden aangeduid als het E u l e r-gemiddelde $\overline{u_E}$ en de gemiddelde snelheid gemeten wanneer men een bepaald volume-element van de vloeistof in zijn beweging volgt, als het L a g r a n g e-gemiddelde $\overline{u_L}$. Het zoeken naar hun verband moet beschouwd worden als een moeilijk, doch belangrijk probleem in de studie van de turbulente beweging. Dit verband moet als volgt luiden:

$$\overline{u_L} = \overline{u_E} + \tfrac{1}{2}\, d\varphi/dx$$

waarin φ de diffusie-coëfficient van het turbulente veld is.

VI

In de theorie van de diffusie als gevolg van continue onregelmatige bewegingen heeft T a y l o r enige betrekkingen tussen correlatie-functies onderzocht. Ter uitbreiding van zijn werk kunnen wij ook opgeven:

$$R_0^{(2p)}(\eta) = (-1)^p \frac{\overline{u^{(p)2}}}{\overline{u^2}} R_p(\eta)$$

waarin: R_0 = de correlatiecoëfficient voor u:

$$R_0(\eta) = \frac{\overline{u(t)u(t+\eta)}}{\overline{u_2}},$$

$R_0^{(2p)}$ = de afgeleide van $R_0(\eta)$ t.o.v. η van de orde $2p$,

R_p = de correlatiecoëfficient voor $d^p u/dt^p$.

G. I. Taylor, „Diffusion by continuous movements", *Proc. London Math. Soc.* (2), **20**, 196–212, (1922).

VII

In tegenstelling tot de vorming van loess in Europa door het water, is de loess in Noord-China meegevoerd door de wind van uit Mongolië en het bekken van Lop Nor, hoewel tussen Noord-China en die gebieden hoge bergen gelegen zijn.

VIII

De Gobi onderscheidt zich van andere woestijnen onder meer doordat de overheersende bewerker van erosie in de Gobi het water is en niet de wind.

IX

Het klimaat van de Gobi vertoont een zekere fluctuatie, waarbij de tegenwoordige tijd een periode van toenemende droogte is, althans tot voor kort was.

X

De oude Chinese geschriften over klimaatveranderingen in China vanaf ca. 2000 jaar geleden wijzen op een vrij nauw verband van de klimaatfluctuaties met de perioden van de zonnevlekken. Een periode van ongeveer 300 jaren uit deze twee pulsaties afgeleid, kan ook aangetoond worden uit de boomgroei van verschillende tijden.

XI

In Zuid-Californië zijn experimenten uitgevoerd over het verband tussen ontbossing en waterafvoer. Daarbij is gevonden dat de waterafvoer ten dienste van watervoorziening en bevloeiing, door ontbossing kan worden bevorderd. De toepassing hiervan is echter niet altijd juist en ongevaarlijk.

XII

De Amerikaanse hulp in de economische opbouw van China is niet altijd zonder gevaar.

XIII

Een democratische coördinatie in de programma's van wederopbouw in de verschillende landen is noodzakelijk om conflicten te voorkomen.

XIV

De wetenschappelijke werkers moeten mede strijden tegen het opkomen van een totalitaire macht, om zich niet te laten gebruiken ten behoeve van doeleinden die zij niet kunnen aanvaarden.

XV

Uit de fysiologie en de leer der evolutie van den mens volgen afdoende bewijzen tegen het dogma van ras-superioriteit.

XVI

Bloedmenging is goed voor de cultuur van een land.

XVII

Andersson onderstelt dat de cultuur van het roodkleurige aardewerk van China afkomstig is uit het Westen of uit Centraal Azië. Na het bekend worden van vondsten van een cultuur van zwartkleurig aardewerk heeft deze onderstelling veel van haar waarschijnlijkheid verloren.

J. G. Andersson, „An Early Chinese Culture", Bull. Geolog. Survey of China, no. 5, part I, 1–68, (1923).